Ferdinand Gerhard Wiechmann

Sugar Analysis

For Refineries, Sugar-Houses, Experimental Stations

Ferdinand Gerhard Wiechmann

Sugar Analysis
For Refineries, Sugar-Houses, Experimental Stations

ISBN/EAN: 9783337139605

Printed in Europe, USA, Canada, Australia, Japan

Cover: Foto ©Andreas Hilbeck / pixelio.de

More available books at **www.hansebooks.com**

FOR REFINERIES, SUGAR-HOUSES, EXPERIMENTAL STATIONS, ETC.,

AND AS A

HANDBOOK OF INSTRUCTION IN SCHOOLS OF CHEMICAL TECHNOLOGY.

BY

FERDINAND G. WIECHMANN, Ph.D.,

Instructor in Chemical Physics and Chemical Philosophy, School of Mines,
Columbia College ;
Consulting Chemist to the Havemeyers and Elder Sugar Refining Company,
Brooklyn, N. Y.

NEW YORK:

JOHN WILEY & SONS,

53 East Tenth Street.

1890.

ROBERT DRUMMOND,
Electrotyper,
444 & 446 Pearl Street,
New York.

FERRIS BROS.,
Printers,
326 Pearl Street,
New York.

PREFACE.

It has been the aim of the writer to prepare a concise yet thorough treatise on Sugar Analysis that should prove of service to the practising chemist as well as to the student of this branch of analytical chemistry.

Within the past few years numerous changes have been made in the older methods of sugar-analysis, new methods have been devised, and many researches of importance to sugar-chemistry have been accomplished.

The current literature of the day devoted to sugar and its interests, abounds in matter pertinent to the subject. A great number of these investigations have, however, appeared only in foreign journals and have therefore not been accessible to all; moreover, they occur scattered through so many different publications that a critical study of the same involves no inconsiderable outlay of time and labor.

A work that should give a general survey of this field seemed therefore both desirable and timely, and it has been with the aim indicated in view, that this publication was undertaken.

The greatest difficulty encountered was the making of a proper choice from the wealth of material at hand.

The schemes selected and here offered, embrace those methods of analysis which, after careful investigation, and, in many cases, after prolonged trial in practice, have seemed to the writer best adapted to the requirements of a technical laboratory.

iii

A glance at the Table of Contents will show at once the plan and scope of this manual.

Instead of taking up for discussion, as is usually done, the different products met with in sugar-laboratories, such as raw sugars, refined sugars, liquors, molasses, etc., and describing for each in turn the determination of their constituents, it has been deemed more expedient to discuss the methods of determining the individual constituents, as sucrose, invert-sugar, water, ash, etc., independently of the products in which they may occur, and then to add such comments and suggestions as certain contingencies would seem to call for.

By the adoption of this plan numerous repetitions have been avoided.

Wherever feasible, examples have been inserted in the text to aid in the understanding of the principles discussed, and of the calculations explained.

Numerous references are given throughout; these will, it is hoped, incite to a study of the original memoirs.

The tables have been selected with the greatest of care, prompted by a desire to introduce only the most accurate. To ensure uniformity of basis, several of these tables have been calculated expressly for this issue. The publication of the formulæ by which the different tables were obtained, should prove a welcome feature to the student.

A list of books and of periodical literature bearing on Sugar Analysis is appended. Asterisks attached to titles show that the publications so marked were consulted in the preparation of these pages, and indicate the obligations of

SCHOOL OF MINES,   THE AUTHOR.
COLUMBIA COLLEGE, 1890.

TABLE OF CONTENTS.

CHAPTER I.

CHAPTER II.

CHAPTER III.

CHAPTER IV.

CHAPTER V.

CHAPTER VI.

CHAPTER VII.

CHAPTER VIII.

CHAPTER IX.

SUGAR ANALYSIS.

CHAPTER I.

POLARIZATION — POLARISCOPES — HYDROMETERS — FLASKS —
THERMOMETERS—BALANCES—WEIGHTS.

Polarization.—If a ray of light strikes a glass mirror
and makes an angle of about 55° with the normal of the
mirror, the ray is not only reflected, but is endowed with
certain properties, and is said to be polarized.

In Fig. 1, *ab* is the incident ray, *bc* the polarized ray.
A plane conceived as passed through *abc* is called the
plane of polarization.

If a polarized ray is allowed to fall upon a
second mirror, parallel to the first, it is again
reflected at the angle above mentioned. If this
second mirror is turned around *bc*, its inclina-
tion to the horizontal being preserved unchanged, the intensity of the reflected ray

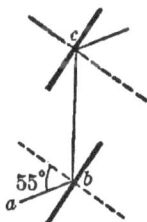

Fig. 1.

continuously diminishes until, when the rotation has
been carried through 90°, the light is extinguished completely. If the rotation be carried beyond this point the
mirror becomes again illumined; and when it has been
turned through 180°, the reflection is again at its maximum of brightness. In other words, the intensity of the
reflected light is greatest when the incident ray and the

1

polarized ray, after reflection from the second mirror, are in the same plane, and least, when these rays are in planes at right angles to each other.

Polarization of light can also be produced by other means: by repeated single refractions, or by double refraction in certain crystals—Iceland-spar, for instance.

If a plate of quartz, cut at right angles to its principal axis, is inserted between two mirrors placed as above described, and traversed by a polarized ray, the image of the quartz will appear in color in the upper mirror. The color of the image changes with the turning of the mirror; the order in which the colors appear is the same as found in the solar spectrum: red, yellow, green, blue, and violet.

This phenomenon is termed circular polarization. It depends on the property possessed by quartz of rotating to a different degree the planes of polarization of the various colored rays which compose white light. One variety of quartz shows these colors in the order named when the mirror is turned to the right; a second variety of the mineral exhibits the colors in this sequence only when the rotation of the mirror is to the left. These varieties of quartz are respectively termed right-rotating and left-rotating, or dextrogyrate and lævogyrate.

Among other bodies which share with quartz the property of circular polarization are the sugars when in solution. Some of the sugars are dextro-rotatory: for instance, sucrose, dextrose, and raffinose; others rotate the plane of polarized light to the left, as lævulose and sorbinose.

The extent to which the plane of polarized light is turned by quartz, by a sugar solution, or any other optically active substance, depends on the thickness of the layer which the polarized ray has to traverse. The thicker the plate or the longer the column of solution, the greater the rotation of the ray. Whereas in the case of a quartz-plate the thickness of the plate is the only factor to be considered, in sugar solutions the concentration of the solution, i.e., the amount of sugar in the solution, must be taken into account.

Polariscopes.—Basing on this property of circular polarization, instruments have been constructed by which the strength of solutions containing optically active substances can be determined. They are called polariscopes or polarimeters. Polariscopes intended for general scientific work are provided with a circular disk, graduated in such a manner that the angle of rotation can be conveniently read. Instruments intended for some special purpose, as, for instance, for sugar analysis, are generally provided with a scale which, if certain directions have been followed in the preparation of the solution, will at once indicate in percentage the amount of the optically active substance present. Polariscopes designed especially for sugar analysis are termed saccharimeters.

The principle on which these instruments are constructed is briefly this: A ray of light is polarized by passing through a prism, called the polarizer, and generally made of Iceland-spar; the ray is then made to traverse a column of sugar solution of known length. Emerging from this, it passes through a second prism of Iceland-spar, the analyzer, which corresponds to the second mirror in the apparatus previously described. It now only remains to ascertain the extent to which the plane of polarized light has been rotated by the sugar solution. The arrangements by which this is effected differ in the various forms of saccharimeters, but in the more modern instruments it is generally accomplished by allowing the light on its emergence from the analyzer to pass through a layer of quartz, the thickness of which (capable of accurate measurement) can be so regulated as to exactly compensate the rotation produced by the sugar solution. It is assumed that the rotatory dispersion of sugar corresponds to that of quartz.

The field of vision of a saccharimeter is either one of color, or else exhibits, when correctly set at zero, a uniform faint tint; polariscopes showing the latter are known as half-shade instruments, and can be used by color-blind persons, as well as by others.

The arrangement of the optical parts of a saccharimeter is shown in the accompanying Figs. 2 and 3.

Fig. 2.

Soleil-Ventzke-Scheibler Polariscope.

1. Magnifying-glass for reading scale.
2. Telescope for observing field of vision.
3. Nicol prism, analyzer.
4. Quartz-wedge, fixed, bearing vernier.
5. Quartz-wedge, movable, bearing scale.   } Rotation
6. Quartz-plate. { Dextro-rotatory if 4 and 5 are lævo-rotatory.   Compensator.
                 { Lævo-rotatory if 4 and 5 are dextro-rotatory. }
7. Double quartz-plate (dextro- and lævo-rotatory).
8. Nicol prism, polarizer.
9. Quartz-plate, dextro- or lævo-rotatory. } Regulator.
10. Nicol prism.

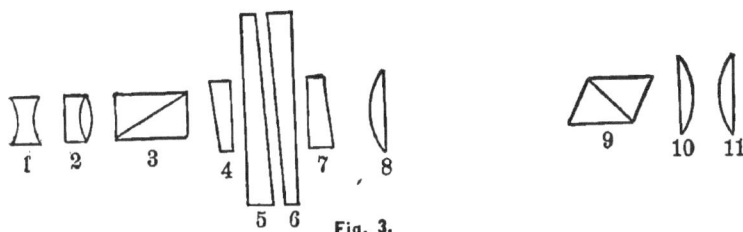

Fig. 3.

Double-wedge Compensator Polariscope, Schmidt and Haensch Construction.

1. Eye-piece.
2. Objective.
3. Nicol prism, analyzer.
4. Quartz-wedge. ⎫
5. Quartz-wedge. ⎬ Constituting the Double-
6. Quartz-wedge. ⎪ wedge Compensator.
7. Quartz-wedge. ⎭
8. Lens.
9. Nicol prism.
10. Lens.
11. Lens.

The scales of saccharimeters are constructed by ascertaining the number of degrees, minutes, and seconds which a definite amount by weight of pure sugar dissolved in water and made up to 100 cubic centimetres will rotate the polarized ray. This is marked as 100, and the scale is then divided into one hundred parts.

If the same weight of an impure sugar is brought into solution and polarized under the same conditions, the reading in the polariscope of course at once expresses percentage of the active substance present.

The scales of different saccharimeters have their 100 mark correspond to different weights of pure sugar. In the Duboscq instrument it is 16.192 grammes, in Wild's apparatus it is 40.000 grammes, and in the Ventzke-Soleil 26.048 grammes. These values are termed the "normal weights" of the respective instruments.

Equivalence in Degrees of Different Saccharimeters.

Grammes of Sugar
in 100 Cubic Centimetres.

1° scale of Mitscherlich	= 0.750
1°. " " Soleil-Duboscq	= 0.1619
1° " " Ventzke-Soleil	= 0.26048
1° " " Wild (sugar scale)	= 0.100
1° " " Laurent and Duboscq (shadow)	= 0.1619

One-degree on the Scale of—	Corresponds to—	Corresponds to—	Corresponds to—	Corresponds to—
	Mitscherlich.	Soleil-Ventzke.	Soleil-Duboscq.	Wild .
Mitscherlich...........		2.879°	4.635°	
Soleil-Ventzke..........	0.346°		1.608°	2.648° ; (
Soleil-Duboscq	0.215°	0.620°		1.619°
Wild (sugar scale)	0.133°	0.384°	0.618°	

Equivalence in Circular Degrees.

J 1° Soleil-Duboscq	= 0.2167 circular degree	D.
J 1° " "	= 0.2450 " "	J.
J 1° Soleil-Ventzke	= 0.3455 " "	D.
J 1° " "	= 0.3906 " "	J.

The letters J and D represent certain rays of light. The former signifies the mean yellow or transition tint, the latter the sodium ray. The amount of rotation which the plane of polarization experiences, called the angle of rotation, varies with the wave-length of the ray: it is least for the red, and greatest for the violet ray.

In saccharimeters using white light (gas or lamp), this value is generally given for the transition-tint, which means the color complementary to mean yellow light.

In order to adjust a polariscope, first obtain by the telescope a sharp and clearly-defined view of the field.

Then turn the screw attached to the quartz-wedge until both halves of the field are, in color instruments, of the same tint; or if the polariscope is a half-shade apparatus, until both halves of the field are equally illumined.

If the instrument is provided with double-wedge compensation, the red scale is first set exactly at zero, and the manipulation is then carried out as described above.

When this has been done the position of the scale is carefully read through the magnifying-glass. The zero of the scale should be exactly in line with the zero mark on the vernier; if this is not the case, they must be brought into the required position by a slight turning of the screw-micrometer provided for the purpose. Care must be taken that the screw in connection with the analyzer be not mistaken for the other screw, or the whole apparatus will be thrown out of order.

If it is impossible to obtain a uniform shade or tint on both sides of the centre line of the field, the polarizer and the analyzer must be brought into adjustment.

This is done by removing the movable and the stationary quartz-wedges, as well as the compensation quartz-plate; the cover is then closed, and the key having been inserted in the screw-head connected with the analyzer (this screw-head is generally placed on the right-hand side of the polariscope), the key is turned until the tint in both halves of the field is uniform.

The wedges and the plate which had been removed are then replaced, and the zero-point accurately adjusted.

When the instrument has been correctly set at zero,

a quartz-plate of known value, preferably one approximating the average test of the sugar solutions to be examined, is inserted in the instrument, and the correctness of that part of the scale ascertained.

The zero-point should be determined before every observation; where press of work renders this impracticable, the observation should be insisted on at least twice daily—in the morning before a polarization is made, and again in the middle of the day.

When a solution is introduced for reading, the telescope must first be properly focussed, as before stated, to insure a clear and sharply defined view of the field.

If the scale stood at zero before the tube filled with the solution was introduced, a glance through the glass will after its introduction show the halves of the field to be of different colors; or, if a half-shade polariscope is used, one half of the field will appear dark and the other light.

The screw attached to the quartz-wedge is then turned until equality in tint or shade shall have been restored to the whole field.

It then only remains to read the scale. Most instruments have the degrees divided into tenths. First it must be determined how many whole degrees the zero of the scale is removed from the zero of the vernier. When this has been ascertained, attention must be given to the tenths of a degree indicated. The number of divisions marking tenths on the vernier are counted until one is found which coincides perfectly with a division on the movable scale, that is to say, which appears to form a continuation of that line. This division repre-

sents the number of tenths indicated. The accompany·
ing figure, for instance, shows 30.7 degrees.

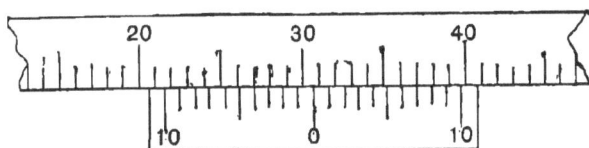

Fig. 4.

The sources of error in saccharimeters are numerous
and therefore every instrument before being placed in use,
should be carefully examined.

The principal difficulties that may be encountered are
the following:

The scale may be too long or too short. Adjust the
zero-point exactly. Make 100 c.c. of a sugar solution by
dissolving the normal weight of chemically pure sugar*
in water, and polarize. This solution should read 100
degrees (per cent) on the scale if the instrument is correct.
If it does not read 100, the instrument should be rejected.

The scale may be right in some places, and wrong in
others. This is the case when the surfaces of the quartz·
wedges are not perfectly plane. In half-shade polari·
scopes provided with double compensation wedges, this
cannot occur, as any inequality would be noticed at
once. In other polariscopes, the scale may be examined
by pure sugar solutions of different densities, by means
of the "control tube" of Schmidt and Haensch, or by
quartz-plates.

The following figures, taken from a table calculated
by Schmitz, show the number of grammes of pure sugar
which must be made up to 100 c.c. aqueous solution in

---

* For preparation of chemically pure sugar see page 17.

order to show the corresponding degree on a polariscope
having 26.048 grammes for its normal weight:

Polariscope Degrees.	Grammes C. P. Sugar in 100 c.c. solution.	Polariscope Degrees.	Grammes C. P. Sugar in 100 c.c. solution.	Polariscope Degrees.	Grammes C. P. Sugar in 100 c.c. solution.
1	0.260	35	9.0,7	69	17.954
2	0.519	36	9.357	70	18.216
3	0.779	37	9.618	71	18.476
4	1.039	38	9.878	72	18.738
5	1.298	39	10.138	73	18.998
6	1.558	40	10.398	74	19.259
7	1.817	41	10.659	75	19.519
8	2.078	42	10.919	76	19.781
9	2.337	43	11.180	77	20.042
10	2.597	44	11.440	78	20.302
11	2.857	45	11.701	79	20.564
12	3.117	46	11.961	80	20.824
13	3.376	47	12.222	81	21.085
14	3.637	48	12.482	82	21.346
15	3.896	49	12.743	83	21.608
16	4.156	50	13.003	84	21.868
17	4.416	51	13.264	85	22.130
18	4.676	52	13.524	86	22.391
19	4.936	53	13.784	87	22.652
20	5.196	54	14.044	88	22.912
21	5.456	55	14.305	89	23.174
22	5.716	56	14.566	90	23.435
23	5.976	57	14.826	91	23.696
24	6.236	58	15.087	92	23.957
25	6.496	59	15.347	93	24.219
26	6.756	60	15.608	94	24.480
27	7.016	61	15.868	95	24.742
28	7.276	62	16.130.	96	25.002
29	7.536	63	16.390	97	25.265
30	7.796	64	16.651	98	25.525
31	8.056	65	16.912	99	25.787
32	8.315	66	17.173	100	26.048
33	8.577	67	17.433		
34	8.837	68	17.694		

This method of testing requires a separate solution
for each degree of the scale which is to be examined.

If the weights necessary to this mode of examination
are not available, the tests can be made by dissolving the
normal weight of chemically pure sugar in different volumes of water at the normal temperature. Thus with a

German saccharimeter 26.048 grammes of such sugar will, when dissolved—

in 100 c.c. water polarize 100.00 degrees.
" 105 " " " 95.23 "
" 110 " " " 90.90 "
" 115 " " " 86.95 "
" 120 " " " 83.33 "

If a control-tube is used, but few solutions are needed, as this tube is so arranged that it can be lengthened or shortened at will. A funnel receives the superfluous solution when the tube is shortened, and a scale attached, shows the length of the column in millimetres. A simple calculation gives the reading which will be shown by the polariscope if this is correct.

If quartz-plates are used to test the accuracy of different parts of the scale, care must be taken that the surfaces of the plates are perfectly plane, that they are inserted in the optical axis of the instrument and at right angles to it.

The quartz-plates themselves should, before being used to control polariscopes, be examined as to their accuracy. One of the ways of ascertaining their value, that is to say, the amount by which they rotate a plane of polarized light, is to measure their thickness.*

This measurement is effected most accurately by means of a spherometer. This consists of a movable screw supported in the centre of three arms, upon which the apparatus rests. The screw is provided at its lower end with a steel point; near its upper end there is fastened a circular plate of metal, the circumference of which is divided into several hundred equal divisions. Fastened

---

* Open to objections, because the specific rotatory power of quartz is not a constant value. Zeitschrift des Vereines für Rübenzucker-Industrie. Vol.

to one of the supporting arms is a metal bar, also bearing a graduation; its graduated edge is placed at right angles to the circular disk.

Parallel to the latter, and attached to the bar, is a sliding-scale which can be set and fastened at any desired height. The graduation of the sliding-scale is so made, that nine of its divisions correspond to ten divisions on the disk.

When the thickness of a plate of quartz, for instance, is to be measured, the screw is first adjusted in such a manner that it shall just touch the perfectly level surface on which the apparatus has been placed.

The sliding-scale is next fastened on the bar exactly on a level with the circular disk.

Suppose the latter to bear five hundred equal divisions, and the graduated bar to be divided into halves of a millimetre. The threads of the screw are so cut that one complete revolution of the screw, indicated by the graduated disk fastened to it, raises the screw through one half of a millimetre. To effect the measurement the screw is first raised sufficiently so as to allow the quartz-plate to be slipped beneath it; when this has been done, the screw is carefully lowered until contact is secured between its point and the quartz-plate. From the number of revolutions through which the screw has been turned, the thickness of the quartz-plate is determined; with a spherometer graduated as here assumed, the measurement will be exact to the one ten-thousandth part of a millimetre.

Besides giving attention to the points already referred to, care must be taken that the Nicol prisms and the lenses are not dusty, and that the illumination is perfect.

The light must be steady and of an unvarying intensity, as the field of vision is materially affected by the flicker-ing of the flame. The end of the instrument must not be placed too near the light, as the heat affects the cement which holds the prisms in position.

The polariscope-tubes must be of exactly the pre-scribed length, as the amount of deviation of the polarized ray produced by an optically active substance depends, among other conditions, on the length of the column of the substance which it traverses. The length of tubes can readily be determined by measuring them with a metal rod made of the standard length. The ends of the pol-arization-tubes must be ground perfectly plane-parallel.

Another point to be borne in mind is the fact that the glass covers of the polarization-tubes may be optically active, either by nature of the glass, by being screwed down too tight, or by not having both surfaces perfectly parallel. The latter difficulty can be readily recognized by taking a glass cover between two fingers and rotating it rapidly, at the same time looking through it at some fixed object. If the latter seems to be moving, the glass is not plane-parallel, and should be rejected.

Hydrometers.—The hydrometers used in the analysis of saccharine solutions embrace specific-gravity hydrome-ters and instruments graduated according to an arbitrary scale. To the latter belong the Baumé hydrometers, and the Brix or Balling spindles. The degrees of a Brix hy-drometer indicate percentage by weight of sugar, when immersed in a solution of pure sugar.

The suggestion has been made to replace the Baumé scale by a scale graduated in the so-called densimetric degrees.

These values are found by taking the specific gravity corresponding to any given Baumé degree, ignoring the unit, and dividing the decimals by 100.

Example.—

Baumé Degrees.	Densities.	Densimetric Degrees.
0	1.0000	0.00
5	1.0356	3.56
10	1.0731	7.31
50	1.5161	51.61

This scale has, however, not yet been adopted in general practice.

The range of scale in each and all of these hydrometers of course varies greatly, according to the ideas and preference of the makers, and of those who use the instruments. The following will be found to be convenient graduations for the ordinary requirements of refinery and laboratory:

Specific-gravity Scale.—Range from 1.095 to 1.106. The scale bears twelve full divisions, and these are divided into halves. Temperature of graduation, 17°.5 C.

The Brix Hydrometers.—Range from 0° to 28°, and covering three instruments: the first from 0° to 8°, the second from 8° to 16°, the third from 16° to 28°. Each degree is divided into tenths.

The Baumé Hydrometers for Liquids heavier than Water.—For general use in the refinery, a scale on a single instrument ranging from 0° to 50°, and divided into quarters or halves, will prove sufficient. For work at the "blow-ups" the range of scale is from 27° to 32°, and each degree is divided into tenths. For the syrup-boiler a scale from 32° or from 38° to 44°, also divided into tenths, is desirable. For laboratory work the range is

from 0° to 45°, best carried over three or more instruments: for instance, from 0° to 20°, from 20° to 35°, and
from 35° to 45°; the subdivision to be in tenths of a
degree.

It is a matter of great importance that the hydrometers used in analytical work be correct. Every instrument should be examined in at least three places, these
being preferably chosen at points corresponding to the
upper, the middle, and the lower part of the scale.

If a correct instrument is at hand (ascertained to be
correct by careful examination), other hydrometers of
the same scale are readily tested by comparison with the
standard hydrometer. If a standard is not available,
the testing must be done in comparison with very accurate specific-gravity determinations, made by a balance.
If the instrument tested is a specific-gravity hydrometer,
the balance determinations are of course directly compared
with its readings; if it is a Brix or a Baumé spindle, the
corresponding specific-gravity values can be ascertained
from Table I.

Methods of Testing Hydrometers.—METHOD I.—The
balance determinations are made by weighing first a
specific-gravity flask or pyknometer,* perfectly clean and
dry. The flask is then filled with distilled water at the
temperature at which the hydrometer was graduated.
This had best be 17°.5 C., and if the hydrometers are
made to order, this temperature should be insisted on for
the graduation.

The weight of the flask filled with water up to the
mark is next taken. A solution is then prepared by dis-

* The neck where the mark is placed, should be narrow, and the flask
should have a tightly-fitting stopper to prevent loss by evaporation.

solving pure sugar in water. The density of this solution is such that it corresponds approximately to one of the points marked on the scale of the hydrometer which is being tested. The temperature of the solution is made to correspond exactly with the temperature at which the specific-gravity flask was previously filled, and the weight of this flask now filled with the sugar solution is accurately determined.

Subtracting the weight of the flask from these two weighings gives respectively the weight of equal volumes of water and of sugar solution. Dividing the latter value by the former, gives the specific gravity of the sugar solution.

Example.—

Weight of specific-gravity flask + water,	40.0403
" " " " "	15.0811
Weight of water in flask,	24.9592
Weight of specific-gravity flask + sugar solution,	42.5810
" " " " "	15.0811
Weight of sugar solution in flask,	27.4999

$$27.4999 \div 24.9592 = 1.1018$$
$$\text{Specific gravity of sugar solution} = 1.1018$$

Some of the sugar solution is poured into a glass cylinder, the temperature carefully brought to 17°.5 C., and the hydrometer, perfectly clean and dry, inserted. It should be allowed to glide down slowly into the solution in order that no more of the stem shall be immersed than necessary. Care must also be taken that the instrument floats free, that is, does not come into contact with the sides.

When the hydrometer has come to rest, a reading of the scale is made and compared with the specific gravity obtained by the balance. The indications of specific-gravity hydrometers should of course agree exactly with the balance determinations; for Brix and for Baumé instruments the limit of agreement should be placed at ± 0°.15. The cheaper Baumé hydrometers, ranging from 0° to 50°, will, however, rarely agree closer than ± 0°.25, and this degree of accuracy will suffice for the practical working purposes of the refinery.

METHOD II.—If the hydrometer is a specific-gravity hydrometer of limited range, it may be tested by immersion in solutions of chemically pure sugar; these solutions are prepared as follows:*

Sp. Gravity.	Grammes C. P. Sugar.	Grammes distilled Water at 17°.5 C.
1.095	22.6	77.4
1.097	23.0	77.0
1.100	23.7	76.3
1.103	24.3	75.7
1.106	25.0	75.0

METHOD III.—If a balance is not available, the testing of specific-gravity hydrometers may be accomplished by the aid of a polariscope. This method is also applicable to Brix and to Baumé hydrometers if their degrees are translated into the corresponding specific-gravity values.

Prepare pure sugar by washing best granulated or powdered block-sugar repeatedly with an 85 per cent alcohol. The washing should be done with a volume of alcohol equal to from three to five times the volume of

---

* Based on the table given in Stammer's Lehrbuch der Zuckerfabrikation, 2d edition, p. 26 et scq.

the sugar. The washed sugar must then be perfectly dried at the temperature of about 100° C., and kept in an air-tight jar. A solution of this sugar is made, the temperature taken, and the hydrometer inserted in it with all the care and precautions previously referred to. After the reading of the hydrometer has been noted, the solution is polarized, and the polarization is multiplied by the factor (Table IV) corresponding to the specific gravity of the solution, corrected, if necessary, for temperature (Table II). If the hydrometer is correct (of course a correct polariscope is premised), the result of the multiplication of the polarization by the factor must be 100.

Example.—

Specific gravity of solution corrected
for temperature, ... 1.096
Factor, ... 1.042
Polarization, ... 96.0
96.0 × 1.042 = 100.0.

Graduation of Flasks.—Two methods are used. The first, the scientifically correct one, is to graduate in *true* cubic centimetres. A true cubic centimetre represents the space occupied by 1 gramme of water weighed in vacuo at a temperature of 4° C.

The second method, known as Mohr's, omits the reduction to volume at 4° C. and to weight in vacuo.

METHOD I.—To graduate a flask at any given temperature, ascertain from Table XVII the weight of 1 cubic centimetre of water at that temperature. Then correct for weighing in air, that is to say, reduce the weighing in air to weighing in vacuo by assuming each gramme of water weighed in air to be 1 milligramme too light.* Tare the flask accurately, place the correct weights on one scale-pan, and weigh the corresponding weight of water into the flask.

Example.—To graduate a flask to hold exactly 100 cubic centimetres at 15° C. Table XVII shows that 1 cubic centimetre of water at 15° C. weighs 0.99916 grammes.

Hence $100 \times 0.99916 = 99.916$ grammes.

As the weighing is to be made in air, to reduce to weighing in vacuo,

$$99.916 \times 0.001 = 0.099916$$

must be subtracted from the former figure:

$$99.916000$$
$$0.099916$$
$$\overline{99.816084}$$

Therefore 99.8161 grammes of water at the temperature of 15° C. must be weighed into the flask.

METHOD II.—The required number of grammes of water (at the temperature chosen) corresponding to the desired volume in cubic centimetres are weighed into the flask, and the resulting volume marked on the flask. These "cubic centimetres" are of course larger than the true cubic centimetres.

Example.—To graduate a flask to hold 50 cubic centimetres at 15° C., 50 grammes of water at 15° C. are weighed into the flask, and the volume occupied is marked as 50 c.c.

Verification of Graduated Glass Vessels, in true Cubic Centimetres.—Fill to the mark with distilled water of

---

* This presupposes the use of brass weights. If the weight of water exceeds 100 grammes, 1.06 milligrammes instead of 1.00 milligramme must be taken in above calculation.

the temperature at which the vessel was graduated, and weigh.

Add to this weight 1 milligramme for each gramme of water weighed.

The density of the water at the temperature of the experiment is to be found in Table XVII.

If P = Corrected weight of the water,

Q = Density of water at temperature of the experiment relative to water at 4° C.,

t = Temperature of the water in the experiment;

then the volume in cubic centimetres contained in the vessel at the temperature $t°$ is

$$V = \frac{P}{Q}.$$

Example.—A flask holds 50.072 grammes of water at 15° C.

The weight in vacuo will be   50.072

                      +   0.050

                      50.122 grammes,

and the capacity at 15° C. will be

$$\frac{50.122}{0.99916} = 50.16 \text{ cubic centimetres.}$$

Thermometers.—The thermometers should be, if possible, compared with some standard instrument. This applies especially to the thermometer which is to be used to determine the temperature while ascertaining the polarization of inverted sugar solutions. It will answer to verify, on Centigrade thermometers intended for ordinary use, the zero and the 100 mark; on a Fahrenheit instrument, the 32° and the 212° mark; and to see that the degrees are of equal size.

The zero-mark on the Centigrade scale (32° Fahrenheit) is ascertained by placing the bulb and part of the stem in snow or pounded ice for about a quarter of an hour. The vessel in which the snow or ice is placed should be provided with a small opening at the bottom, through which the water is drained off as it is formed.

To obtain the 100° C. (212° F.) mark, the thermometer is suspended in the vapor of boiling water, care being taken that it does not dip into the water. The pressure of the atmosphere should be 760 mm. at the time; if not, a correction for the variation must be made.

The reading of one scale can be translated into that of the other by the following formulæ:

$$C = \frac{5}{9}(F - 32)$$

$$F = \frac{9}{5}C + 32$$

For a comparison of the different thermometric scales see Table XVIII.

Balances.—For weighing out samples for polarization, a balance capable of weighing up to 300 grammes and sensible to 1 milligramme will answer. For water and ash determinations an analytical balance should be used; this should be sensible to 0.1 of a milligramme, and be capable of bearing a charge up to 200 grammes.

A good balance* should give the same result in successive weighings of the same body; the two halves of

---

* See Deschanel-Everett : Natural Philosophy.

the beam should be of equal length; it should be sensible to a small load, and it should work quickly.

It is an easy matter to determine whether a balance possesses these properties. Repeated weighings of the same load will quickly establish whether the balance is consistent with itself; this depends principally on the trueness of its knife-edges.

To determine whether both halves of the beam are of the same length, the two pans should be loaded with equal weights. If the arms are of unequal length, the pan attached to the longer arm will descend.

To test the sensibility, load both pans with the maximum weight which they are intended to bear, and then add to one of the pans the weight to the extent of which the balance is supposed to be sensible. The addition of this slight extra weight should cause the pan on which it has been placed, to descend.

Weights.—The weights used, both the regular weights for analytical purposes, and the so-called sugar-weights (normal and half normal), should be verified from time to time, as they will in daily use unavoidably suffer some wear and tear. Most of the weights are so made that the plug or stopper unscrews from the body of the weight, and slight deficiencies in weight can readily be corrected by inserting tin-foil or small shot into the cavity after removing the plug.

Should the weights be too heavy, a little filing will readily remedy the evil.

CHAPTER II.

Sampling Sugars and Molasses.—Too much importance cannot be attached to the securing of correct samples, that is to say, to the obtainment of samples which
shall be representative of the substance examined.

The samples of raw sugar are drawn with a long steel
bar resembling the half of a pipe cut longitudinally.
A hole having been made in the package, the "tryer,"
as it is called, is inserted, rotated completely, and then
withdrawn. The sample which fills the hollow in the
tryer is removed and is placed in a can.

When syrups or molasses are to be sampled, a rod or a
stick is inserted in the bung-hole of the barrel and rapidly
withdrawn ; the adhering liquid is placed in a can, and
the operation repeated until sufficient has been obtained.

When sugars in hogsheads are sampled, the hogshead is placed on its side. The manner of inserting the
tryer differs. The Government takes its sample by run-
ning straight through the contents from centre to centre
of the heads; at some refineries the tryer is run through
diagonally from head to head.

Melados are sampled through the bunghole of the
hogshead.

In a refinery, 100 per cent of all sugars, syrups, and
molasses are sampled.

23

The U. S. Government varies its requirements as to the number of packages to be sampled, with the nature of the package:

Of hogsheads, tierces, boxes, and barrels, 25 per cent are required for sample and 100 per cent for a resample; of centrifugals and of beet-sugars, in bags, 5 per cent for sample and 5 per cent for resample; of mats, $2\frac{1}{2}$ per cent for sample and $2\frac{1}{2}$ per cent for resample; of baskets, 10 per cent for sample and 10 per cent for resample; of "Jaggeries," Pernambuco, and Brazil sugars, 5 per cent for sample and 5 per cent for resample.

When the samples have been taken and are brought to the laboratory for analysis, it is necessary, either to make a separate analysis of every mark in a lot, or, as this is generally not feasible, to prepare a representative sample.

In order to do this, fix upon some definite quantity by weight as the unit weight. Weigh out this amount, proportionate to the number of hogsheads in each mark, and place in a well-closed jar.

For example, suppose a lot of sugar contained four marks, A, B, C, and D.

$$\text{Mark A} = 1000 \text{ hogsheads,}$$
$$\text{`` } \text{B} = 200 \text{ ``}$$
$$\text{`` } \text{C} = 350 \text{ ``}$$
$$\text{`` } \text{D} = 70 \text{ ``}$$

Then take from:

$$\text{A} = 100 \text{ grammes}$$
$$\text{B} = 20 \text{ ``}$$
$$\text{C} = 35 \text{ ``}$$
$$\text{D} = 7 \text{ ``}$$

For analysis, if necessary, crush the sample, thoroughly mix the contents of the jar, and then proceed as usual.

As some lots come in mixed packages, that is to say, partially in hogsheads, bags, tierces, and barrels, a certain relation between these has been assumed; it is as follows:

1 hogshead = 2 tierces.
"   = 8 barrels.
"   = 8 bags.

To prepare average samples of refined sugars, proceed in a similar manner, as directed above.

Determination of Color of Sugar and Sugar Solutions.—The color-tests made on sugars and on sugar solutions are generally only comparative, that is to say, the color of the sample examined is compared with that of some other sample which is taken as the standard.

In the examination for color of raw sugar, the so-called "Dutch standards" are usually employed. These consist in fifteen samples of raw sugar, numbered from No. 6 to No. 20, and ranging in color from a dark-brown (No. 6) to almost a white (No. 20). They are prepared and sealed with great care by a certain firm in Holland. The samples are renewed every year, and serve as standards for the twelve months following their issue.

In examining the color of sugar solutions, to learn, for instance, how effectively a certain sugar has been decolorized in passing through bone-black, two test-tubes, beakers, or cylinders made of white glass. are filled to an equal height with, respectively, the sample under examination and the "standard" solution with which the sample is to be compared, both solutions of course being of equal density.

Various forms of apparatus have been designed for effecting color comparison. In some, the "standard" solution is replaced by colored-glass disks of tints ranging from a pure white to a dark yellow or brown; by combination of these it is possible to produce almost any shade desired.

The colorimeter probably most used is that of Stammer. As the depth of color of a solution is proportional to the length of a column of such solution, there is ascertained in this instrument the height of a column of the solution which will in color correspond to the tint of a "standard" colored-glass disk inserted in an adjoining tube. The scale is graduated in millimetres. If, for instance, a depth of one millimetre of the solution corresponds to the normal tint, the color is said to be 100. If two millimetres depth of solution are required to match the tint, the color is 50; if four millimetres, 25; and so on.

Determination of the Density of Solutions.—*By the Specific-gravity Flask.*—The most accurate way to determine the density (specific gravity) of a solution is by means of a specific-gravity flask (pyknometer) and a delicate balance, as already described on page 15. The weight of the flask, empty and dry, having been ascertained, and the weight of distilled water which it will hold at 4° C. or at the temperature at which it was graduated being known, once for all, it is only necessary to fill the clean and dry flask exactly up to the mark with the solution whose specific gravity is to be determined. If the solution has not been brought to the temperature at which the flask was graduated, before the flask is filled with it, this must certainly be done before the weighing is made, in order that the weight of equal volumes of the water and the solution may be obtained.

The flask filled with the solution is weighed, the weight of the flask subtracted from this figure, and the remainder divided by the weight of the corresponding volume of water. The result is the specific gravity of the solution.

By Pipette and Beaker.—An adaptation of the method just described, and which is convenient for rapid working, is the following:

A pipette capable of holding a certain volume, say 10 or 20 c.c., is placed in a glass beaker; both pipette and beaker of course must be perfectly clean and dry. The combined weight of the two is taken and noted.

The pipette is then filled with distilled water at the temperature which is to be made the normal temperature, —preferably 17°.5 C. The pipette is replaced in the beaker, and the combined weight of the pipette, beaker, and water is determined. The vessels having been again cleaned and dried, the solution whose specific gravity is to be determined, is brought to the standard temperature, and the pipette filled with it up to the mark. The weight of pipette, beaker, and solution is then determined. The calculation to be made is exactly as before explained, the combined weight of beaker and pipette taking the place of the weight of the pyknometer in the previous method.

By Hydrometers.—The hydrometer selected for making the determination may be a specific-gravity hydrometer or an instrument graduated according to an arbitrary scale (Brix, Baumé).

Whenever a solution is to be tested, care must be taken to have it as free of air-bubbles as possible. If the solution whose density is to be determined is a thick syrup or a molasses, it had best be poured into a vessel provided at the bottom with a stop-cock. This vessel may advantageously be enclosed in a water-jacket. This can be heated and the molasses thus readily warmed, which will greatly hasten and facilitate the rising of the air-bubbles. When they have all risen to the top, the liquid is drawn off from below, without disturbing the frothy layer on the surface.

The liquid is placed into a glass cylinder, which must stand perfectly level, and the hydrometer is carefully and slowly inserted. It must float free in the liquid, that is, it must not be permitted to touch the sides of the cylinder. When the hydrometer has come to rest, the point up to which it is immersed in the solution is read and recorded. The temperature of the solution is determined, and if not of the standard temperature, a correction therefor must be made. (See Table II or III).

The readings of the specific-gravity, the Brix, and the Baumé hydrometers can each readily be translated into the terms of the others by Table I.

By Glass Spheres.—For approximate density determination small glass balls of different weights are sometimes used. A number engraved or etched on each, designates the density of a liquid in which it will float.

Beginning with the heavier, the balls are successively thrown into the solution whose density is to be determined, until a ball is found which will float in the liquid tested. The number engraved on this ball indicates the density of the solution. Of course regard must here also be had to the temperature of the liquid.

By Mohr's Hydrostatic Balance.—From one end of the beam of this balance a glass bob, preferably one provided with an accurate thermometer, is suspended by a fine platinum wire. The other end of the beam is provided with a counterpoise to the bob; this counterpoise terminates in a fine metal point, and serves as the tongue of the balance. It shows the beam to be in equilibrium when the same remains at rest in a horizontal position directly opposite to a fixed metal point.

The balance, when correctly adjusted, is in perfect equilibrium when the glass bob hangs freely suspended in air.

That part of the beam between the fulcrum and the end from which the bob is pendant, is provided with nine graduations, numbered from one to nine. Accompanying the balance are five weights or riders. The largest two are each equal to that weight of distilled water (at a certain temperature, usually 15° C. or 17°.5 C.), which the glass bob displaces when it is immersed. The other three riders weigh respectively one tenth, one hundredth, and one thousandth as much as the large rider.

When the bob is immersed in water, one of the large riders must be placed at that end of the beam from which the bob is suspended. This will restore the equilibrium, and the balance then indicates the specific gravity 1.000.

If the bob is immersed in a liquid heavier than water, this liquid having been brought to the temperature for which the balance was graduated, some of the other riders also must be placed on the beam in order to restore the equilibrium. The position of these riders indicates the specific gravity of the solution, each rider according to its weight, representing respectively as many tenths, hundredths, or thousandths as is expressed by the numbered division on the beam where it is placed.

Determination of Alkalinity.—The alkalinity of the different products of a refinery may be caused by potassium, by sodium, by lime, or even partially by free ammonia. It has, however, become customary to report the alkalinity in terms of calcium oxide (caustic lime).

Alkalinity is determined by the addition of an acid of known strength to a known weight or volume of the product examined, until neutrality has been established.

The acid used may be either sulphuric, nitric, or hydrochloric acid, the first of these being the one most commonly employed. As indicator, litmus solution, phenol-phthalein, or rosolic acid (corallin) is available.

Litmus turns red with free acid, while phenol-phthalein is colorless, and rosolic acid * is colorless or shows a pale yellow color with free acid. The indications afforded by these agents are said to be not identical, and any set of comparative determinations therefore should be carried out with the *same* indicator, whichever of these may be selected.

The acid used is generally of "tenth-normal" strength. To prepare this there are needed of:

Sulphuric oxide   4.00 grammes SO, in 1 litre of water.
Hydrochloric acid 3.637   "   HCl   "   "   "   "
Nitric acid   6.289   "   HNO, "   "   "   "

The acid should be delivered from a burette divided into tenths of a cubic centimetre.

To effect an alkalinity determination, 10 to 20

---

* Use alcohol for dissolving. Of phenol-phthalein, 1 part in 500 parts of alcohol, of rosolic acid, one 1 part in 100 parts of alcohol of 90%.

grammes of the product to be tested are weighed out and dissolved, or, if a solution is to be examined, from 10 to 20 cubic centimetres are measured out and placed in a porcelain dish. A few drops of the indicator having been added, the acid is allowed to flow in from a burette until the change in color of the indicator shows the reaction to be finished.

1 cubic centimetre of $\frac{N}{10}$ (tenth normal) sulphuric acid corresponds to 0.0040 gramme sulphuric oxide, 0.0028 gramme calcium oxide, or 0.0047 gramme potassium oxide.

The number of cubic centimetres of acid used, multiplied by 0.0028, show therefore the amount of calcium oxide present.

Example.—25 cubic centimetres of a sugar solution (specific gravity 1.198) required 2.4 cubic centimetres $\frac{N}{10}$ sulphuric acid to effect neutralization. This represents 0.0028 × 2.4 = 0.00672 gramme calcium oxide.

$$25.0 : 0.00672 :: 100 : x.$$

$x =$ 0.02688 per cent calcium oxide. This is *percentage by volume*. If *percentage by weight* is required, the above value must be divided by the specific gravity of the solution, or if a specific-gravity determination and this subsequent calculation are to be avoided, the solution to be tested must be in the first place weighed out, and not measured.

Determination of Acidity.—To determine the acidity of a solution, syrup, molasses, etc., the same course is followed as above described, only of course the solution added to effect neutralization is one of sodium hydrate (caustic soda), potassium hydrate (caustic potash), or calcium hydrate (slaked lime), and the change of

color of the indicator, if litmus, must be from red to blue,
or if phenol-phthalein or rosolic acid are employed, from
colorless to a bright crimson.   Of these solutions the calcium hydrate is least desirable, as the carbonic acid of the
atmosphere readily precipitates in it calcium carbonate,
and so changes the strength of the solution.   A $\frac{N}{10}$ sodium-
hydrate solution contains 3.996 grammes NaOH in 1 litre
of water.

Test for Sulphurous Oxide in Sugar.—Dissolve from
10 to 20 grammes of the sugar in about 25 cubic centimetres of distilled water.   Pour into a flask, and add
about 5 grammes of chemically pure zinc (free from
sulphur), and 5 cubic centimetres of chemically pure hydrochloric acid.   Suspend a paper moistened with acetate
of lead solution in the neck of the flask.   If sulphur
dioxide is present, it will be liberated from its combinations and changed into sulphuretted hydrogen, and this
gas will turn the acetate of lead on the paper a brown or
a black color, owing to the formation of sulphide of lead.

CHAPTER III.

SUCROSE : IN THE ABSENCE OF OTHER OPTICALLY ACTIVE
SUBSTANCES.

Optical Analysis.—METHOD I. *With Balance.*—Weigh
out 26.048 grammes of the sample.* Dissolve in 50 to
75 c.c. of water, and pour into a 100 c.c. flask. Add basic
acetate of lead solution,† the amount depending on the
nature of the sugar tested, and then add a few drops of
a solution of sodium sulphate to insure the precipitation
of any excess of the lead salt.‡
Filter rapidly into a covered beaker to avoid concentration of solution by evaporation ; rejecting the first few
drops entirely, fill the 200 mm. polarization-tube, and
take the reading. Several readings should be taken on
the same solution, and their mean recorded.

* The sample must previously have been well mixed; if the sugar, as is
frequently the case, contains lumps, the whole sample must be thoroughly
crushed before the mixing.

In cold weather sample-cans brought in from out-of-doors, should be
allowed to stand in the laboratory until their contents shall have approximately attained the temperature of the room. This is done in order to
avoid condensation of moisture on the cold sugar, as this would slightly
lower the polarization.

† *Basic Acetate of Lead.*—To 300 grammes acetate of lead and 100
grammes litharge (oxide of lead) add 1 litre of water. Allow to stand for
twelve hours in a warm place, with occasional stirring; then filter, and
preserve in a well-closed bottle.

The basic acetate of lead must show a strongly alkaline reaction, and
have a specific gravity ranging from 1.20 to 1.25 at a temperature of
17°.5 C.

‡ It is impossible to prescribe the quantity of the basic acetate of lead
solution to be used; always, however, employ the least amount that will
produce the desired effect, tor a voluminous precipitate causes an error in
polarization.

33

With very dark sugars and with syrups, the half-normal weight, 13.024 grammes, is often taken, dissolved up to 100 c.c., and the reading made in a 200 mm. tube; or the normal weight is used, and the reading effected in the 100 mm. tube.

It must be remembered that the temperature exerts an influence on the polarization reading. The colder the solution the higher the reading; a variation in temperature of two degrees Centigrade,* is stated to cause a difference of one tenth of a degree on the polariscope.

Decolorization of dark solutions is effected by adding to the solution some bone-black dust previously prepared,† by use of the so-called Gawalowsky's decolorizer, or by " blood carbon." Whichever of these is employed, the least amount possible should be used.

For very dark sugars and molasses the use of sodium sulphite (a 10 per cent solution) and basic acetate of lead is recommended.‡ The sodium sulphite is first introduced, about 2 c.c., and then the basic acetate of lead solution is gradually added with constant shaking, till no further precipitation occurs. If necessary, the filtrate from this can be subjected to the action of sulphurous acid and bone-black.

Opalescence or a slight but persistent turbidity of the solution to be polarized, can be overcome by the addition of a little " alumina cream."§ Three to five cubic centi-

---

* Die Deutsche Zuckerindustrie, vol. xiv. p. 503.

† Warm for several hours with hydrochloric acid to dissolve the phosphate and carbonate of lime; then wash with boiling water till all traces of chlorine are removed ; dry at about 125° C., and keep in a well-closed jar.

‡ Allen : Commercial Organic Analysis, vol. i. p. 201.

§ Precipitate a solution of alum, not too concentrated, by ammonic hydrate. Wash the precipitate until all the salts have been removed, and the washings no longer turn red litmus blue.

metres are ample, if not more than the half-normal weight has been used for making the solution. This reagent is of little value as a decolorizer, but very efficient with high-grade sugars that show the troublesome opalescence.

The saccharimeters now in universal use record the amount of sucrose in per cent, provided the normal weight* of the sample has been used, and the reading has been effected in a 200 mm. tube; if a 100 mm. tube has been used, the reading must be doubled; or if the half-normal weight has been taken, and the polarization has been effected in a 200 mm. tube, the reading must of course also be doubled.

If for any reason the normal or the half-normal weight has not been taken, a simple calculation will serve to figure the percentage of sucrose in the sample. Suppose, for instance, that 9.000 grammes had been weighed for polarization and that these were dissolved up to 50 c.c. A polarization of this solution in a 200 mm. tube = 62.00.

As a rotation of one degree represents 0.13024 gramme sucrose, there are contained in the sample $0.13024 \times 62 = 8.07488$ grammes pure sucrose.

Hence $9.00000 : 8.07488 :: 100 : x.$ $x = 89.72.$

Therefore the sample contains 89.72 per cent sucrose.

A more direct way of figuring this is by means of the formula:

$$\frac{P \times W'}{W} = \text{per cent sucrose.}$$

$P =$ polarization of the solution;

$W' =$ normal or half-normal weight of the instrument used;

$W =$ weight of substance taken for polarization.

---

* The normal weight for the German instruments is 26.048 grammes; for the Duboscq polariscopes it is 16.192 grammes.

$$Example.— \frac{62.0 \times 13.024}{9.0} = 89.72.$$

Results so obtained can be verified by calculating the amount of sugar which would be necessary in order to indicate 100 degrees on the polariscope. This is known as Scheibler's method of "One hundred polarization."

Example.—In the case just discussed, a polarization of 89.7 required 13.024 grammes of the sugar: how much will be required to produce a rotation of 100 degrees on the instrument?

$$89.7 : 13.024 :: 100 : x. \qquad x = 14.5195.$$

Therefore 14.5195 grammes of this sample are polarized in the usual manner, and if they indicate 100 per cent, the result previously obtained, is correct.

Table VII, by Scheibler, obviates the necessity of this calculation, showing at once the amount that must be used.

Method II. *Without Balance.*—The percentage of sucrose in a sample can also be obtained without making a weighing. A solution is made and the specific gravity of the solution is determined, either directly by a specific-gravity hydrometer, or else by some other hydrometer (Brix, Baumé), the readings of which are translated into the corresponding specific gravity (Table I).

The polarization of the solution is then made, and the percentage of sucrose calculated by the formula:

$$S = \frac{P \times .2605}{D},$$

in which S = percentage of sucrose,
P = polarization of the solution,
D = specific gravity.

If the solution needs clarifying, it is placed into a

graduated flask, the amount of basic acetate of lead solution that is added, is noted, and the reading increased in proportion.

Example.—Specific gravity of solution, 1.0909 ; Polarization of solution = 35.0.

To 100 c.c. of solution added 5 c.c. basic acetate of lead solution; this corresponds to 5 per cent of 35.0 = 1.75.

Hence corrected polarization = 36.75 per cent.

$$\frac{36.75 \times .2605}{1.0909} = 8.77 \text{ per cent sucrose.}$$

This calculation can be avoided by consulting Table VI. This table is used in the following manner :

Example.—Corrected specific gravity = 1.0339 ; Polarization = 25.0.

In a line with the specific gravity 1.0339, and in the horizontal column marked 2, is found the number .504 This multiplied by 10 = 5.040.

In a line with the specific gravity 1.0339, and in the column marked 5, is found the number 1.260.

Adding these values, 5.040
1.260

Percentage of sucrose = 6.300

The simple polarization of a sugar, syrup, liquor, magma, or sweet-water shows the percentage of sucrose in the sample as it is. Sometimes, however, it is necessary to know what this percentage would be if the water in the sample were removed ; in other words, it may be desirable to ascertain the percentage of sucrose in the "dry substance."

The percentage of pure sugar in the "dry substance" is referred to as:

The **Quotient** of **Purity**, or **Exponent.**—There are several ways of determining this. The most accurate method undoubtedly, but also the one demanding most time, is the following:

METHOD I.—Determine polarization of the normal weight of the sample as previously described (p. 33). Determine the percentage of water by drying to constant weight (see p. 76). Subtract the percentage of water from 100, and divide the remainder into the polarization multiplied by 100.

Example.—Polarization of syrup,   33.00;
            Water in syrup, per cent,   24.16.

$$100.00$$
$$\underline{24.16} \qquad 3300 \div 75.84 = 43.5$$
$$75.84$$

Polarization on dry substance = 43.5.

METHOD II.—Determine polarization of the normal weight of the sample as previously described (p. 33). Determine the degree Brix of the sample. Correct for temperature (Table III).

Calculate polarization on the dry substance by the formula: $\dfrac{\text{Pol.} \times 100}{\text{Degree Brix}}$.

Example.—Polarization,   40.00;
            Density,   50° Brix at 24° C.;
            Correction for temperature, + 0.49
            Degree Brix corrected for temperature,
            — 50.49.

100.00 ÷ 50.49 = 1.9806, factor ;

40.00 × 1.9806 = 79.22, polarization on the dry substance, or coefficient of purity.

METHOD III. *Ventzke's Method.*—Prepare a solution of the sugar which shall have the specific gravity 1.100 at 17°.5 C. Take the reading of this solution in a 200 mm. tube. This polariscope reading shows at once the percentage of *pure* sugar in the dry substance. This is the case, because a solution made by dissolving 26.048 grammes of chemically pure sugar in water up to 100 c.c. has the specific gravity of 1.1000 at the temperature of 17°.5 C., and a column of this solution 200 mm. in length, indicates 100 per cent in the German polariscopes.

The following table prepared by Gerlach* shows the specific gravity of the above solution at the temperatures given:

Temperature. ° C.	Specific Gravity.	Temperature. ° C.	Specific Gravity.	Temperature. ° C.	Specific Gravity.
0	1.10324	16.5	1.10028	23	1.09834
5	1.10266	17	1.10014	24	1.09802
10	1.10192	17.5	1.10000	25	1.09770
11	1.10168	18	1.09986	26	1.09736
12	1.10144	18.5	1.09972	27	1.09702
13	1.10119	19	1.09957	28	1.09669
14	1.10095	19.5	1.09943	29	1.09635
15	1.10071	20	1.09929	30	1.09601
15.5	1.10057	21	1.09897		
16	1.10043	22	1.09865		

As the preparation of a solution which is to have

* Jahresbericht über die Untersuchungen und Fortschritte auf dem Gesammtgebiete der Zuckerfabrikation, 1863, p. 234.

a certain specific gravity at a certain temperature is apt to prove a tedious operation, the following modification of Ventzke's method will prove serviceable:

If the temperature at which the solution is prepared is not the normal temperature, a correction must be made (Table. II).

This correction must be subtracted from the reading of the specific-gravity hydrometer if the temperature is lower than the normal, and added, if it is above the normal temperature.

The polarization obtained in the 200 mm. tube must then be multiplied by the factor corresponding to the corrected specific gravity (Table IV).

METHOD IV. *Casamajor's Method.*—Determine the specific gravity or the degree Brix of the solution. Correct for temperature if necessary (Table III). Determine the polarization of this solution and multiply the polarization by the factor corresponding to the degree Brix (Table V).

Example.—Polarization of solution = 61.2;

Brix, = 15°.5 at 22° C.;

Correction for temperature, + 0.31

Corrected degree Brix = 15.81;

Factor corresponding to 15°.8 Brix is 1.548

$61.2 \times 1.548 = 94.74$, which is the polarization on the dry substance, the coefficient of purity.

The quotient of purity obtained by Method I (where the percentage of water is obtained by actual drying out), is called the "true" quotient of purity; if hydrometers are resorted to, as in Methods II, III, and IV, the resulting coefficient is called the "apparent" quotient of purity.

If a syrup or a molasses has been analyzed, the re-

sults of the analysis can easily be calculated into equivalents on the dry substance in the following manner:

The reciprocal of the degree Brix (that is, the quotient obtained by dividing 100 by the degree Brix), gives
a factor by which the percentage of sugar, invert sugar,
and ash must be multiplied in order to reduce them to
the basis of dry substance.

Example.—A syrup of 80°.4 Brix shows on analysis:

'Polarization, 31.2 ;
Invert sugar, 12.5 ;
Ash, 6.0.

100 ÷ 80.4 = 1.2437.

On Dry Substance.

Hence : Polarization, 31.2 × 1.2437 = 38.80 per cent.

Invert-sugar, 12.5 × 1.2437 = 15.55 "

Ash, 6.0 × 1.2437 = 7.46 "

Non-ascertained (by difference) = 38.19 "

100.00 per cent.

If sucrose has to be determined in a molasses, a syrup,
or in sweet-water, the calculation of the result to dry substance can be avoided by aid of Table VIII.

This table has been calculated for use with the German polariscopes (normal weight 26.048 grammes). It
presupposes the addition of 10 per cent by volume of
basic acetate of lead to the sucrose solution examined, and
in its preparation the variable specific rotatory power of
sucrose has also been taken into account.

The use of the table is very simple.

Example.—Density of a sugar solution, 22°.0 Brix.
Polarization (after using 10 per cent by volume of basic
acetate of lead solution for clarifying), 60.3.

In column headed 22°.0 Brix, and opposite to the

number 60 in the column headed "Polariscope degrees," we find 15.72 per cent sucrose. Then turning on the same page to the division for tenths of a degree, in the section headed " Per cent Brix from 11.5 to 22.5," there is given opposite to 0.3 Brix the value 0.08 per cent sucrose.

Hence   60°.0 = 15.72 per cent.

$$0°.3 = 0.08 \quad \text{``}$$

$$60°.3 = 15.80 \text{ per cent sucrose.}$$

Gravimetric Analysis.—Weigh out 13.024 grammes of the sample. Dissolve with about 75 c.c. of water in a 100 c.c. flask. Add 5 c.c. hydrochloric acid containing 38 per cent HCl (sp. gr. 1.188). Heat quickly, in two or three minutes, on a water-bath up to between 67° and 70° C. Then keep at this temperature (as close to 69° C. as possible) for five minutes, with constant agitation. Cool quickly; make up to 100 c.c. Remove 50 c.c. by a pipette, place in a litre flask, and fill up to 1000 c.c. Of this solution take 25 c.c. (corresponding to 0.1628 gramme of sample), neutralize all free acid present by about 25 c.c. of a solution of sodium carbonate prepared by dissolving 1.7 grammes crystallized sodium carbonate in 1000 c.c. of water. Then add 50 c.c. of Fehling's solution, heat to boiling as directed in invert-sugar determination, boil for three minutes, and proceed as directed on page 69.

Calculation.—In Table XI seek the number of milligrammes of copper which agree most closely with the amount of copper found. The corresponding number in the column to the left, shows at once the number of milligrammes of sucrose.

Example.—25 c.c. of the inverted solution = 0.1628 gramme of sample, yielded 0.1628 gramme copper.

This corresponds to 0.082 gramme sucrose; hence there are present in the sample 50.4 per cent sucrose.

As invert-sugar, dextrose, and even raffiinose (after inversion by acid), reduce Fehling's solution, a correction of the results yielded by this method must be made, whenever appreciable quantities of the substances named are present.

If the sample analyzed contains invert-sugar, the amount of this substance multiplied by 0.95 must be subtracted from the "Total sucrose" found, in order to obtain the actual amount of sucrose present. This factor 0.95 is used, because sucrose on inversion yields invert-ugar in the proportion of 100 : 95.

CHAPTER IV.

SUCROSE: IN THE PRESENCE OF OTHER OPTICALLY ACTIVE SUBSTANCES.

THE determination of sucrose can be effected by means of the polariscope, as described in the previous chapter, provided no other optically active bodies are present.

Such substances, however, occur frequently; they may be dextro- or lævo-rotatory. If the presence of such substances is suspected, it will be necessary to perform an inversion by acid, and determine the polarization of the inverted solution.

If no other optically active substances are present besides the sucrose, the polarization before and after inversion will be equal.

If the polarization after inversion is higher than the polarization before inversion, lævo-rotatory bodies are present; if the polarization after inversion is lower than the polarization before inversion, dextro-rotatory substances are indicated.

In the former case invert-sugar, lævulose, etc., must be considered; in the latter, dextrose, raffinose, etc., will have to be looked for.

Clerget's Inversion Method. — Weigh out 26.048 grammes of the sample, and determine the polarization. Of the filtrate, take 50 c.c. for inversion, or weigh out separately 13.024 grammes of the sample.* Dissolve with about 75 c.c. of water in a 100 c.c. flask; add, while agi-

---

* Herzfeld's modification. Zeitschrift des Vereines für Rübenzucker-Industrie, 1888, p. 709.

44

tating the solution, 5 c.c. hydrochloric acid (sp. gr. 1.188), containing 38 per cent HCl. Heat quickly, in two or three minutes, on a water-bath up to between 67° and 70° C. Then keep the temperature of the solution for five minutes as close to 69° C. as possible. Agitate constantly. Then cool quickly, fill with distilled water up to the 100 c.c. mark, and polarize in a tube provided with an accurate thermometer.* The temperature at which the reading is taken should be 20° C.

For dark solutions, molasses, etc., take 26.048 grammes of the sample, dissolve, add basic acetate of lead and sodium sulphate, and fill up to 100 c.c. Filter. Of the filtrate remove 50 c.c. with a pipette, place in a 100 c.c. flask, add 25 c.c. of water, and 5 c.c. of hydrochloric acid containing 38 per cent HCl, and proceed as directed above. The result is calculated by means of the formula:

$$R = \frac{100S}{142.66 - \frac{1}{2}t}.$$

R = sucrose; S = sum of the two polarizations before and after inversion, the minus sign being neglected; t = temperature in degrees Centigrade at which the polarization after inversion is observed.

Example.—Polarization of normal weight before inversion, 87.5;

Polarization of half-normal weight after inversion, − 14.3 at 20° C.

$$\frac{-14.3 \times 2}{-28.6} \qquad \frac{87.5}{28.6} \qquad R = \frac{100 \times 116.1}{142.66 - 10}$$

$$\overline{116.1}$$

$$R = \frac{11610}{132.66} = 87.5$$

* Thermometers constructed expressly for this purpose, and on which the degrees are divided into tenths, are made by C. Haack in Jena, Germany.

It is best to carry out the determination at 20° C. if possible. If, however, the determination is made at any other temperature from 10° C. to 30° C., Table X gives a series of factors by which it is necessary to multiply the difference of the indications, before and after inversion. Of course the factor corresponding to the temperature at which the reading of the inverted solution was made, must be used.

Example.—Direct polarization, 86.0 ;
Polarization after inversion, − 25.0, at a temperature of 22° C.
86.0 + 25.0 = 111.0.

Referring to Table X, opposite to 22° C. there will be found the factor 0.7595. Multiplying 111 × .7595 = 84.3; this is the desired result.

If any other weight than 13.024 grammes is used for the determination, the formula $R = \dfrac{100.S}{142.66 - \frac{1}{2}t}$ does not give quite correct results, because the specific rotatory power of an invert-sugar solution varies also with the degree of concentration of the solution.

Sucrose in the Presence of Raffinose.*—Prepare 26.048 grms. of the sample for polarization, as directed p. 33, and polarize. Of the polarized solution (from which all lead should first have been removed) take 50 c.c. Place in a 100 c.c. flask ; add 5 c.c. concentrated hydrochloric acid (38.8 per cent HCl) and about 20 c.c. of distilled water. Heat on a water-bath up to between 67°

---

* Method prescribed by the German Government to regulate the duty on sugar, July 9, 1887. Several methods and numerous modifications have been proposed to effect the determination of raffinose. For the benefit of those desiring more information on the subject, a list of references is given on the opposite page.

and 68° C. This should take about five minutes. When this temperature has been reached, it should be maintained for five minutes more. The solution is then quickly cooled to 20° C., made up to the 100 c.c. mark, and polarized at exactly 20° C. in a tube provided with a very sensitive and accurate thermometer. This tube should be enclosed in another tube or should be placed in a trough which is filled with water, so that the temperature of 20° C. may obtain throughout the observation.

Author.	Publication.	Year.	Volume.	Page.
Pellet and Biard.	Journal des fabr. de sucre.	1885		
Von Lippmann.	Deutsche Zuckerindustrie.	1885	X.	310
Tollens.	Zeitschrift d. V. f. Rüben-zucker-Ind.	1886	XXXVI.	236
Scheibler.	Neue Zeitschrift f. Rüben-zucker-Ind.	1886	XVII.	233
Creydt.	Zeitschrift d. V. f. Rüben-zucker-Ind.	1887	XXXVII.	153
Creydt.	Zeitschrift d. V. f. Rüben-zucker-Ind.	1888	XXXVIII.	979
Directions of the German Government.	Neue Zeitschrift f. Rüben-zucker-Ind.	1888	XXI.	132
Gunning.	Neue Zeitschrift f. Rüben-zucker-Ind.	1888	XXI.	335
Lotman.	Chemiker Zeitung.	1888	XII.	391
Breyer.	" "	1889	XIII.	559
Schulz.	Zeitschrift d. V. f. Rüben-zucker-Ind.	1889	XXXIX.	673
Wortman.	Zeitschrift d. V. f. Rüben-zucker-Ind.	1889	XXXIX.	767
Lindet.	The Sugar Cane.	1889	XXI.	542
Herzfeld.	Zeitschrift d. V. f. Rüben-zucker-Ind.	1890	XL.	165
Courtonne.	Journal des fabr. de sucre.	1890	XXXI.	

The sucrose and raffinose are calculated by the formulæ :*

$$S = \frac{(0.5188 \times P) - I}{0.845};$$

$$R = \frac{P - S}{1.85};$$

S = sucrose ;

R = raffinose ;

P = polarization of normal weight (26.048 grms.) before inversion ;

I = polarization of normal weight (26.048 grms.) after inversion.

Example.—Polarization before inversion, 93.8
　　　　　Polarization after inversion, — 12.7

$$93.8 \times 0.5188 = + 48.66344$$
$$- 12.7 \times 2 \quad = - 25.40000$$
$$\overline{+ 74.06344}$$

$74.06344 \div 0.845 = 87.6.$　$S = 87.6$ per cent.

$$93.8$$
$$- 87.6$$
$$\overline{6.2} \div 1.85 = 3.35.$$　$R = 3.35$ per cent.

If the observation of the inverted raffinose solution has not been made at 20° C. a correction of 0.0038° for each degree Centigrade above or below 20° C. must be

---

* Tollens and Herzfeld prefer to calculate these values by the formulæ:
$$S = \frac{(0.5124 \times P) - I}{0.000} \quad \text{and} \quad R = \frac{P - S}{1.852}.$$

introduced. This correction is effected by the formula :*

$$\left.\begin{array}{c}\text{Polarization}\\\text{after inversion}\\\text{at } 20° \text{ C.}\end{array}\right\} = \left\{\begin{array}{c}\text{Polarization}\\\text{after inversion}\\\text{at } t° \text{ C.}\end{array}\right\} + 0.0038\ S(20 - t),$$

in which S represents the sum of the polarizations before and after inversion.

Example.—Suppose a solution of sucrose and raffinose polarized :

before inversion, 105°.0 ;

After inversion, −22°.0 at a temperature of 18°.2 C.

Then the polarization after inversion at 20° C. will be equal to :

$$- 22.0 + 0.0038(105.0 + 22.0)(20. - 18.2)$$
$$- 22.0 + 0.0038(+ 127.0)(+ 1.8)$$
$$- 22.0 + 0.86868$$
$$= - 21.13.$$

Sucrose in Presence of Dextrose (Glucose). *Qualitative Tests.*—A number of tests have been proposed for the qualitative examination of a sugar for dextrose. Among these the following are possibly the most serviceable :† Thoroughly dry the sample to be examined. Prepare a solution of methylic alcohol saturated with dextrose.‡ Pour some of this solution on the dried sample, and stir for about two minutes. Allow the residue to settle, and pour off the clear solution. Repeat this treatment. If any dextrose is present, some chalky-

---

* Zeitschrift des Vereines für Rübenzucker-Industrie, vol xl. p. 201.

† Casamajor, Journal of the American Chemical Society, vol. ii. p. 428, and vol. iii. p. 87.

‡ 100 c.c. methylic alcohol, showing 50° by Gay-Lassac's alcoholometer, dissolve 57 grammes of dry glucose. The specific gravity of the solution is 1.25.

white particles and a fine deposit will remain, for dextrose
is practically insoluble in the solution employed, while
the sucrose will go into solution.

The test is best made in a beaker with a flat bottom
or on a pane of glass.

If a syrup is to be examined for the presence of dextrose, provided the dextrose has been added in sufficiently large quantity, and the syrup has the usual density of about 40° Baumé, the following test may be
applied: The direct polarization of the syrup should
show a percentage of sugar not higher than the number
of Baumé degrees which indicate the density. If, for
instance, a syrup of 40° Baumé should show a direct
polarization of 55.0, some dextro-rotatory substance, most
probably dextrose, must have been added to this syrup, as
an unadulterated product of this description would be a
mixture of crystals and syrup, and could not be a clear
syrup.

The polariscope may also be resorted to for detecting
the presence of dextrose.

The manner of procedure is simple:

The solution is prepared as usual for the polariscope;
then, immediately after preparing it, a reading is taken;
the solution is allowed to remain in the tube for some
time, and repeated readings are taken at certain intervals. If dextrose is present, the successive readings will
become lower and lower, for dextrose is bi-rotatory.
Readings on the solution are continued until the rotatory
power has become stationary; it may take up to fifteen
hours before this is attained.

When this point has been reached, treatment with
hydrochloric acid (attempted inversion), will produce no

effect, the dextro-rotatory power of the dextrose remaining unchanged.

Quantitative Methods.—The quantitative methods for the determination of dextrose in the presence of sucrose are based either on optical analysis, on gravimetric analysis, or on a combination of both.

Among the methods of the first type, that of hot polarization, due to Drs. Chandler and Ricketts, is probably the best.*

This method depends upon the following well-known facts:

1. *Dextrose*, under the conditions of analysis, exerts a constant effect upon the plane of polarized light at all temperatures under 100° C.

2. *Lævulose.* The action of lævulose is not constant, the amount of rotation to the left being diminished as the temperature is increased.†

3. *Invert-sugar*, being a mixture of one half dextrose and one half lævulose, does not affect the plane of polarized light at a certain temperature, somewhere near 90° C.‡ (for it can easily be seen that the constant dextro-rotatory power of dextrose must be neutralized by the varying lævo-rotatory power of lævulose at some such temperature. The exact temperature is determined by experiment).

4. *Cane-sugar*, when acted on by dilute acids, is converted into invert-sugar, while dextrose remains practically unaltered.

---

* Abstracted from a report made by A. L. Colby to the Chairman of the Sanitary Committee in the Second Annual Report of the State Board of Health of New York, 1882.

† Watts' Dictionary of Chemistry, vol. v. p. 464.

‡ Ibid. p.465.

Hence, if a "mixed sugar" is heated with dilute acids, the cane-sugar present is converted into invert-sugar, which, with that originally present (due to the process of manufacture), is optically inactive at a certain temperature (near 90° C.); while the artificial dextrose, preserving its specific rotatory effect, will at this temperature show a deviation to the right in proportion to the amount present.

It is only necessary, therefore, to secure some means of heating the observation-tube of the ordinary polariscope, so that readings may be taken at any temperature under 100° C. The middle portion of a Soleil-Ventzke saccharimeter, ordinarily intended for the observation-tube alone, is so modified as to admit of the interposition of a metallic water-bath, provided at the ends with metal caps, which contain circular pieces of clear plate-glass. The tube for holding the sugar solution to be polarized, is made of platinum, and provided with a tubule for the insertion of a thermometer into the sugar solution. The metallic caps at the end of the tube rest on project-ing shelves inside the water-bath, thus bringing the tube into the centre of the bath, where it is completely sur-rounded by water. The cover of the water-bath is arranged for the insertion of a thermometer, so that the temperatures of the water-bath and of the sugar solution may both be ascertained. The water-bath is heated from below by two to four small spirit-lamps or gas-burners. The first step in using the instrument is to determine, by experiment, the exact temperature of the sugar solution, at which invert-sugar is optically inactive on polarized light. This will vary slightly with different instruments. For the particular instrument and thermometer used in

the investigations referred to, 86° C. was found, by repeated experiment, to be the temperature of the pure inverted sugar solution at which the reading was *zero* on
the sugar scale.

The next step taken was the determination of the
value of a degree of the scale in terms of the glucose
known to be the variety used to adulterate cane-sugar. It
was found that the rotation to the right at 86° C. was 41°,
when using a solution containing in 100 c.c. fifteen grammes
of a sample containing 85.476 per cent chemically pure
glucose. Hence as fifteen grammes was the amount taken,
$15 \times \frac{85.476}{100} \div 41 \times 100 = 31.2717$ grammes, which represents the amount of chemically pure glucose necessary to
read one hundred divisions on the sugar scale of the instrument used; or, each division $= 0.312717$ grammes chemically pure glucose. (A duplicate determination made, by
using 26.048 grammes, gave as a factor 0.312488.)

The success of the process depends greatly upon the
care exercised in preparing the sugar solution for the
polariscope. The inversion and subsequent clarification
were accomplished as follows :

26.048 grammes of the sugar to be examined were completely dissolved in about 75 c.c. of cold water, and were
treated with 3 c.c. of dilute sulphuric acid (1 to 5 by
volume) on a water-bath at a temperature of about 70°
C. for thirty minutes. The solution thus inverted was
then rapidly cooled, nearly neutralized with sodium carbonate solution (saturated), transferred to a 100 c.c. flask,
and the gummy matters, etc., precipitated with 5 c.c. of a
solution of basic lead acetate.* The flask was then filled

---

* Prepared by boiling for thirty minutes 440 grammes neutral lead acetate with 264 grammes litharge, in one and a half litres of water ; dilut-
ing when cool to two litres, and siphoning off the clear liquid.

to the mark, the solution transferred to a small beaker, mixed with enough bone-black to clarify completely, and then thrown on a fluted filter. The amount of bone-black necessary to effect decolorization depends on the grade of the sugar and on the color of the solution. It was not found necessary to use, even with sugars of the lowest-grade, more than five grammes.*

The clarified inverted sugar solution was then placed in the platinum polarization-tube, the water-bath was filled with cold water, the thermometers were adjusted, and the temperature gradually raised to 86° C. This part of the operation should take about thirty minutes. If the sample is unadulterated, the polariscope reading would be zero at 86° C., while if starch-sugar is present the amount of deviation to the right, in degrees and fractions, multiplied by the proper factor and divided by the amount taken, would give the per centage of chemically pure glucose added as an adulterant.

Gravimetric Method.—The following method is based on gravimetric determinations, and is independent of all optical data. This will be recognized as an advantage when the great influence is remembered that temperature-fluctuations exert on the rotatory power of invert-sugar.

Unfortunately, however, the destruction of the lævulose by hydrochloric acid (Sieben's process), on which this whole scheme of analysis is based, is not always accomplished with the same certainty,† and the results obtained by this method must therefore be received with some caution and reserve.

---

* The bone-black used was pulverized to pass through an 80-mesh sieve, dried at 110° C. for three hours, and kept in a well-closed bottle.

| The Author, School of Mines Quarterly, 1890, vol. xi.

The determinations to be made are:
1. Total sucrose. See p. 42.
2. Total reducing sugars. See p. 69.
3. Dextrose after destruction of the lævulose by Sieben's treatment. See p. 59.
Determination No. 1 embraces:
  a. Invert-sugar formed from the sucrose by inversion.
  b. Invert-sugar existing as such.
  c. Bodenbender's substance (regarded as invert-sugar).
  d. Free dextrose (if present).
Determination No. 2 embraces:
  a. Invert-sugar.
  b. Bodenbender's substance (regarded as invert-sugar).
  c. Free dextrose (if present).
Determination No. 3 embraces:
  a. Dextrose from the inverted sucrose.
  b. Dextrose from invert-sugar.
  c. Dextrose from Bodenbender's substance (regarded as invert-sugar).
  d. Free dextrose (if present).

No. 1 minus No. 2 gives the copper reduced by the (inverted) sucrose. One half of this amount represents the dextrose from this source, i.e., from the sucrose which was turned into invert-sugar.

Subtracting this from No. 3 leaves the copper due to the dextrose of the invert-sugar + the dextrose of Bodenbender's substance (regarded as invert-sugar) + free dextrose, if present. Call this amount x.

If there is no *free* dextrose present, but only invert-sugar and Bodenbender's substance (regarded as invert-sugar), then $2 \times x$ must be equal to the amount of copper found in No. 2.

If there is no invert-sugar, but only sucrose and dextrose, then x will be equal to No. 2.

If there is *free* dextrose present besides the invert-sugar, then $2 \times x$ will be greater than No. 2, and the amount of copper representing the free dextrose will be found, as shown by example No. 3.

Example 1.—Present: sucrose and invert-sugar, but no free dextrose.

Det. No. 1 yields	0.420 Cu
Det. No. 2　"	0.040 Cu
Det. No. 3　"	0.212 Cu
No. 1,	0.420
minus No. 2,	0.040

$$0.380 \div 2 = 0.190 \text{ Cu due to dextrose from the inverted sucrose.}$$

Det. No. 3,	0.212
less	0.190
	0.022

This corresponds to the x above.

$$0.022 \times 2 = 0.044$$
$$\text{Det. No. 2} = 0.040$$

These two values agree within 0.004, and as the limit of difference should be placed at 5 milligrammes of copper, it must be inferred that this solution contained no free dextrose.

Another way of calculating is as follows:

Det. No. 3,	0.212 Cu
Det. No. 1 = 0.420	
less Det. No. 2 = 0.040	
0.380 ÷ 2 = 0.190 Cu	
0.022 Cu	

This is the copper due to the dextrose from the invert-sugar, from Bodenbender's substance (regarded as invert-sugar) and from free dextrose, if any is present.

This amount 0.022 must be equal to one half of No. 2, *if no free* dextrose is present.

No. 2 = 0.040 ÷ 2 = 0.020; hence there is a difference of only 0.002, and therefore there is no free dextrose.

Example 2.—Present: sucrose and dextrose, but no invert-sugar.

Det. No. 1 yields	0.474 Cu
Det. No. 2 "	0.286 Cu
Det. No. 3 "	0.382 Cu

$$\text{Det. No. 1} = 0.474$$
$$\text{less No. 2} = 0.286$$

$$\overline{0.188 \div 2 = 0.094 \text{ Cu}}$$

due to the dextrose of the inverted sucrose.

$$\text{Det. No. 3} = 0.382$$
$$\text{less} \qquad\quad 0.094$$
$$\overline{0.288}$$

This value is not equal to one half of No. 2, but equal to the whole of the copper found in No. 2 (in fact it shows 2 milligrammes of Cu more); hence this solution contained no invert-sugar, but only sucrose and dextrose.

Example 3.—Present: sucrose, dextrose, and invert-sugar.

Det. No. 1,	0.500 Cu
Det. No. 2,	0.300 Cu
Det. No. 3,	0.275 Cu
Det. No. 1,	0.500
less No. 2,	0.300

$$\overline{0.200}$$

.200 ÷ 2 = .100 copper due to dextrose from the inverted sucrose.

$$\begin{array}{ll} \text{No. 3,} & 0.275 \\ \text{less} & 0.100 \\ \hline & 0.175 \end{array}$$

.175 × 2 = 0.350

No. 2 is 0.300; hence, as this value 0.350 is greater than No. 2, yet not twice as great, there must be present invert-sugar and free dextrose. To calculate the amounts respectively of the invert-sugar and of the dextrose, proceed as follows:

No. 2, 0.300 is Cu reduced by the invert-sugar, Bodenbender's substance and dextrose;

0.175 is Cu reduced by one half of the invert-sugar and of Bodenbender's substance, and by the whole of the dextrose;

0.125 × 2 = 0.250 invert-sugar and Bodenbender's substance;

and 0.300 *minus* 0.250 = 0.050 is the Cu reduced by the dextrose.

The 0.250 Cu reduced by the invert-sugar + Bodenbender's substance (regarded as invert-sugar) is equal to 0.1347 invert-sugar.

The 0.050 Cu reduced by the dextrose is equal to 0.0259 dextrose. (Table XV).

The 0.200 Cu reduced by the invert-sugar produced from the sucrose by inversion, corresponds to 0.1015 sucrose; hence the sample contains:

Sucrose, milligrammes, ... 101.5
Invert-sugar (inclusive of Bodenbender's
  substance), milligrammes, ... 134.7
Dextrose, milligrammes, ... 25.9

Knowing the amount of dry substance on which the tests were performed, the calculation to percentage can be readily effected.

Sieben's Process for Destruction of Lævulose.—Take 100 c.c. of a solution made to contain 2.5 grammes on the dry substance of invert-sugar, or of invert-sugar and lævulose, place in a flask, add 60 c.c. of a hydrochloric-acid solution which is six times the strength of a normal solution, and heat the flask for three hours while it is suspended in boiling water. After this has been done, cool immediately, neutralize with a sodium-hydrate solution which is six times the strength of a normal solution, make up to a volume of 250 c.c., and filter. Of the filtrate use 25 c.c. for the determination of the dextrose; this is obtained as follows:

Take 30 c.c. copper-sulphate solution; *
    30 cc. Rochelle-salt solution; †
    60 c.c. water.

Heat to boiling. Add the 25 c.c. dextrose solution, prepared as above, and keep boiling for *two* minutes. Then proceed as with a gravimetric determination of invert-sugar. (See p. 69). Table XV shows the amount of dextrose corresponding to the weight of copper found.

---

* Prepared by dissolving 69.278 grammes C. P. sulphate of copper in distilled water, and making the solution up to 1 litre.

† Prepared by dissolving 173 grammes Rochelle salt, cryst. and 125 grammes potassium hydrate in distilled water, and making the volume up to 500 c.c.

Determination of Sucrose, Dextrose, and Lævulose.

—Several methods have been suggested for the determination of sucrose, dextrose, and lævulose in the presence of each other.

Some of these are combinations of optical and gravimetric methods; as, for instance, those given by Tucker,[*] Apjohn,[†] and Dupré.[‡] The first of these mentioned is directed to the determination of dextrose and lævulose, while the others refer also to the determination of sucrose.

Winter[§] has published an outline of the separation and determination of dextrose and lævulose in the presence of sucrose; his method is based on the action of ammoniacal acetate of lead. This reagent is prepared, immediately before use, by adding ammonic hydrate to basic acetate of lead solution, until the turbidity formed just continues to disappear.

To the solution to be examined, add ammoniacal acetate of lead until no further precipitate is formed. Then filter. The precipitate must be digested with *large* quantities of water, and the washings must be added to the filtrate. This filtrate contains the sucrose.

The precipitate consists of the lead salts of dextrose and lævulose. It is suspended in water, carbonic-acid gas is passed in, and the solution is then filtered.

The filtrate contains the dextrose. This is determined by the polariscope and by its action on alkaline copper solution.

---

[*] Tucker: Manual of Sugar Analysis, 2d Ed., p. 208.
[†] Chemical News, vol. xxi. p. 86; Amer. Reprint, p. 230.
[‡] Loc. cit., p. 97; Amer. Reprint, p. 239.
[§] Zeitschrift des Vereines für Rübenzucker-Industrie, 1888, p. 782.

The precipitate consists of the carbonate and the lævu-losate of lead. This is suspended in water, and sulphu-retted hydrogen gas is passed in. The sulphide of lead is removed by filtration. The filtrate is concentrated by evaporation, and the lævulose is determined by the polari-scope and by its action on alkaline copper solution.

Gravimetric Method.—The gravimetric method de-scribed on page 54 can also be adapted to the deter-mination of sucrose, invert-sugar and dextrose, or lævu-lose. The determinations to be made are the same as those there directed, namely, total sucrose, total reducing sugars, and total dextrose after destruction of the lævu-lose by Sieben's treatment.

The same reserve, however, as there noted, must be exercised with reference to accepting the results ob-tained. Any method by which the destruction of the lævulose could be effected completely and under all cir-cumstances, and leave the dextrose unattacked, would make this method a most valuable one.

The method of calculating the results is analogous to the one before given, and consists of two steps:

Step I. is always the same, and merely establishes whether the dextrose and the lævulose are present in the proportion of 1 to 1, or whether either is in excess.

Step II. determines the amount of this excess, be it of dextrose or of lævulose.

Values determined:

No. 1. Copper reduced by total sucrose + total reducing sugars.

No. 2.   "   "   "   total reducing sugars.

No. 3.   "   "   "   dextrose (after Sieben's treat-ment).

Calculation.

Step I.

No. 1 = Cu reduced by inverted sucrose and total reducing sugars.

Less No. 2 = Cu reduced by total reducing sugars.

Difference = Cu reduced by inverted sucrose. Report the corresponding value as sucrose.

This difference ÷ 2 = Cu reduced by the dextrose of the inverted sucrose. Call this value x.

No. 3 = Cu reduced by the total dextrose (after Sieben's treatment).

Less x = Cu reduced by the dextrose of the inverted sucrose.

Difference = Cu reduced by the dextrose of the total reducing sugars. Call this value y. Then,

$y \times 2 = 2y$ Cu reduced by invert-sugar + free dextrose, if any is present.

Compare this value, $2y$, with No. 2:

If $2y$ = No. 2, invert-sugar *only* is present. If so, report as invert-sugar.

If $2y$ > No. 2, free dextrose is present.

If $2y$ < No. 2, free lævulose is present.

Step II.

When $2y$ > No. 2, free dextrose is present.

No. 2 = Cu reduced by the total reducing sugars.

Less y = Cu reduced by the dextrose from the total reducing sugars.

Difference = Cu reduced by the lævulose of the total reducing sugars. Call this value p.

$p \times 2 = 2p$ Cu reduced by invert-sugar. Report as invert-sugar.

No. 2 = Cu reduced by the total reducing sugars.

less $2p$ = Cu reduced by invert-sugar.

Difference = Cu reduced by the free dextrose.

Step II.

When $2y < $ No. 2, free lævulose is present.

No. 2 = Cu reduced by the total reducing sugars.

Less $2y$ = Cu reduced by the invert-sugar. Report as invert-sugar.

Difference = Cu reduced by the free lævulose.

In these calculations no attention has been paid to the fact that the reducing-power of invert-sugar, dextrose, and lævulose for copper solutions is not identical.

The reducing power of dextrose being considered as 100, that of invert-sugar is 96, and of lævulose 94.

CHAPTER V.

INVERT-SUGAR.

Qualitative Examination for Invert-Sugar.—Test with Methyl-Blue.—Dissolve 1 gramme of methyl-blue in 1 litre of water, and keep for use.

To execute this qualitative test for the presence of invert-sugar, dissolve 20 grammes of the sugar in water, add basic acetate of lead solution, make up to 100 cubic centimetres, and filter. Make the filtrate slightly alkaline with a 10 per cent solution of sodium carbonate, and filter again. Of this filtrate take 50 cubic centimetres, representing about 10 grammes of the sugar tested, place in a porcelain casserole, and add 2 drops of the methyl-blue solution. Then place the casserole over a naked flame, and note accurately when the solution begins to boil.

If the solution is decolorized by boiling, inside of one half-minute, there is sufficient invert-sugar present to permit of a quantitative determination. If it requires from one-half to three minutes boiling to effect disappearance of the blue color, traces of invert-sugar are to be reported; and if decolorization does not take place within three minutes, "no invert-sugar" is recorded.

If the normal weight has been dissolved up to 100 c.c., 20 c.c. of the solution, clarified by basic acetate of lead, are made up to 50 c.c. The lead is removed by adding five drops at a time of the sodium carbonate solution, and the a[...]ion of this reagent, in the same quantity, is continued, until no more precipitation can be detected.

To 25 c.c. of the filtrate one drop of the methyl-blue solution is added; about 10 c.c. of this solution are kept actively boiling over a naked flame for one minute.

If, after thus boiling for one minute, the solution is completely decolorized, it must have contained at least 0.01 per cent of invert-sugar. If it is not decolorized, it contained no invert-sugar, or certainly less than 0.015 per cent.*

Quantitative Determination of Invert-Sugar.—Fehling's solution (Soxhlet's formula) :

Sulphate of copper cryst., 34.639 grms. in 500 c.c. of water.
Rochelle salts, . . . 173.0 grms. in 400 c.c. of water.
Sodic hydrate, . . . 50.0 grms. in 100 c.c. of water.

Keep the sulphate of copper solution in one flask, and the Rochelle-salt-soda solution in another. Mix the two immediately before use. It will be found very convenient to have the solutions in flasks or jars provided with a siphon-arrangement, and to have the delivery-tube so graduated that the required amount may be rapidly, yet accurately measured out. The accompanying figure shows an arrangement answering this purpose.

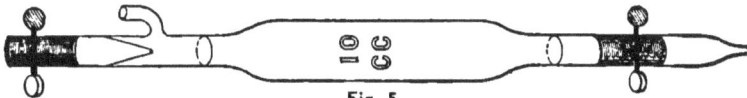

Fig. 5.

Volumetric Methods. Soxhlet's Method.†— Take 25 c.c. of the sulphate of copper solution and add to it 25 c.c. of the Rochelle-salt-soda solution.

---

* Wohl. Zeitschrift des Vereines für Rübenzucker-Industrie, 1888, p. 352.

† Journal für Practische Chemie, New Series, 1880, vol. xxi. p. 227.

Place in à deep porcelain casserole, heat to boiling, and add sugar solution until the fluid, after boiling for two minutes, is no longer blue.

This preliminary test indicates approximately (within about 10 per cent) the amount of invert-sugar present. Next dilute the sugar solution till it contains about 1 per cent of invert-sugar. The true concentration will be 0.9 to 1.1 per cent, which slight deviation from the concentration desired, has no influence on the result.

Take 50 cc. of Fehling's solution, heat, add the requisite amount of sugar solution, boil for two minutes, and then pour the whole solution through a large corrugated filter-paper. Test the filtrate for copper by acetic acid and potassium ferrocyanide.

If copper is found to be present, repeat the test, but take a greater volume of the sugar solution. If the filtrate is found to be free from copper, repeat the test, but take 1 c.c. less of the sugar solution.

Continue with these tests until of two sugar solutions, differing from one another by only 0.1 c.c., the one shows copper, and the other shows no copper in the filtrate. The amount of sugar solution intermediate between these two, must be regarded as the one that will just decompose 50 c.c. of the Fehling solution.

1.0 equivalent of invert-sugar reduces 10.12 equivalents of cupric oxide in solutions made as here prescribed. If the solution be diluted by four volumes of water, 1.0 equivalent of invert-sugar will reduce 9.7 equivalents of cupric oxide.

FEHLING'S METHOD.*—Five, ten, or, if necessary, more

---

* Annalen der Chemie und Pharmacie, 1849, vol. 72, p. 106.

grammes of sugar are weighed out, dissolved in a flask, and the solution made up to 100 c.c. The weight of sugar used varies, of course, with the nature of the sample examined, that is to say, with the amount of invert-sugar it contains. It is advantageous to have the solution of such a strength that 20 c.c. to 50 c.c. will completely precipitate the copper in 10 c.c. of the solution cited above.

The Fehling solution is measured out (using 5 c.c. each of the copper sulphate and the Rochelle-salt-soda solution), placed in a porcelain dish, and quickly brought to the boiling-point. The sugar solution is then run in from a burette (graduated in tenths of a cubic centimetre) until all of the copper in the solution is precipitated as cuprous oxide. The operator is warned of the approach of the end of the reaction by the change in the color of his solution. The blue color disappears and the solution becomes colorless, or, if the sugar solution is colored, assumes a yellow tinge.

The end-point, however, is determined by filtering a few drops of the solution through paper or linen cloth into a very dilute solution of potassic ferrocyanide * and acetic acid. †

If a brownish-red color shows, owing to the formation of cupric ferrocyanide, two tenths c.c. more of the sugar solution are added to the copper liquor, the solution is again boiled, and the test repeated. This is continued until the addition of a few drops of the solution to the ferrocyanide no longer produces the red color.

If a polarization is to be made on the same sample, 19.21 cubic centimetres of the solution for polarization,

* 20 grammes dissolved in 1 litre of water.
† A 10 per cent solution.

prepared by dissolving 26.048 grammes in 100 c.c., and from which the lead has been removed, represents exactly 5 grammes, and may be used for the determination of the invert-sugar. If the French normal weight (16.19 grammes) has been used, 30.8 c.c. are required. These amounts can be measured out from a burette, or pipettes may be procured, graduated to deliver the given volumes of solution.

As 10 c.c. of the copper solution are assumed to correspond to 0.5 gramme of invert-sugar, the calculation is an easy one. If 5 grammes of sugar have been dissolved up to 100 c.c., the reciprocal of the number of cubic centimetres required of this solution to precipitate all of the copper in 10 c.c. of the copper liquor, multiplied by 100, is the direct percentage of invert-sugar sought. (See Table XII.)

Example.—Dissolved 5 grammes of sugar in 100 c.c. Of this solution used 22 c.c. to precipitate all of the copper in the Fehling solution. Referring to Table XII, 22 c.c. will be found to correspond to 4.54 per cent of invert-sugar; hence there is this amount of invert-sugar present in the sample.

DEXTROSE SOLUTION FOR STANDARDIZING THE FEHLING SOLUTION.—Dissolve 4 grammes C. P. anhydrous dextrose, in distilled water, and make up to 1000 c.c. 1 c.c. = 0.004 dextrose.

To test the strength of the copper solution, place 10 c.c. of it in a porcelain dish or casserole, with from 30 to 40 c.c. of water. Boil, and run in the dextrose solution from a burette until all the copper is precipitated.

The number of cubic centimetres of the dextrose solution used, multiplied by 4, represents the number of milligrammes of dextrose required to precipitate the copper in 10 c.c. of the Fehling solution.

Gravimetric Method.—MEISSL-HERZFELD.—Weigh out 26.048 grammes of the sample. Place into a 100 c.c. flask, clarify with basic acetate of lead, make up to 100 c.c., filter, and polarize. Take an aliquot part of the filtrate, add sodium sulphate to remove any lead present, make up to a definite volume, and filter. It is best to arrange the dilution so, that the 50 c.c. of this filtrate, which are to be used for the determination of the invert-sugar, will precipitate between 200 and 300 milligrammes of copper.

To 50 c.c. of the sugar solution prepared as above, add 50 c.c. Fehling's solution (25 c.c. copper sulphate and 25 c.c. of Rochelle-salt-soda solution). Over the wire-gauze above the flame lay a sheet of asbestos provided with a circular opening of about 6.5 cm. diameter; on this place the flask, and arrange the burner in such a manner, that about four minutes are consumed in heating the solution to the boiling-point. From the time that the solution starts to boil—the moment when bubbles arise not only from the centre but also from the sides of the vessel—continue to boil for exactly two minutes with a small flame. Then remove the flask from the flame immediately, and add 100 c.c. of cold distilled water, from which the air has previously been removed by boiling.*

Then filter through an asbestos filter, wash, and reduce to metallic copper.†

---

* The water is added to prevent subsequent precipitation of cuprous oxide.

† This last step is sometimes omitted, the cuprous oxide being weighed after washing and drying, and the corresponding amount of copper calculated.

This operation is carried out in the following manner: Clean thoroughly a small straight calcium-chloride tube, or other tube of similar pattern. Introduce asbestos fibres* so as to fill about half of the bulb. Draw air through while drying, cool, and weigh. Connect with an aspirator, filter the precipitated Cu_2O, wash with hot water, and then, having changed the receiving flask, wash twice with absolute alcohol and twice with ether. Having removed the greater part of the ether by an air-current, connect the upper part of the filter tube by means of a cork and glass tubing with a hydrogen apparatus, and, while the hydrogen gas is flowing through, cautiously heat the precipitate with a small flame whose tip is about 5 cm. below the bulb containing the Cu_2O. The reduction should be completed in from two to three minutes.

After the tube has been cooled in the current of hydrogen, air is once more drawn through and the tube is then weighed.

After an analysis is completed, the asbestos is readily freed from the adhering copper by washing with dilute nitric acid.

The use of the electric current has also been advocated for reducing the precipitate to metallic copper.†

The cuprous oxide is dissolved with 20 c.c. nitric acid (sp. gr. 1.2), the solution is placed into a weighed platinum dish, made up to between 150 and 180 c.c. with

---

* The asbestos must first be prepared by washing successively with a solution of caustic soda (not too concentrated), boiling water, nitric acid, and again with boiling water. When filled into the glass tube the asbestos is made to rest on a perforated platinum cone.

† Formanek Böhm. Ztschr. für Zuckerlndustrie, 1890, vol. xiv. p. 178.

distilled water, and the copper precipitated by the electric current.

The method of calculating the amount of invert-sugar, corresponding to the weight of copper found, can best be illustrated by an example. Suppose that of the 26.048 grammes of sugar dissolved in 100 c.c., 25 c.c. had been removed, clarified with sodium sulphate, made up to 100 c.c., and filtered: 50 c.c. of this filtrate would correspond to 3.256 grammes of substance.

Let this weight be designated by the letter p.

The *approximate* amount of invert-sugar may be assumed to be
$$= \frac{\text{Cu}}{2}.$$

The *approximate percentage* of invert-sugar will be
$$= \frac{\text{Cu}}{2} \times \frac{100}{p}.$$

Representing the former value by Z, the latter by y, we have
$$Z = \frac{\text{Cu}}{2},$$
and
$$y = \frac{\text{Cu}}{2} \times \frac{100}{p}.$$

The ratio between the invert-sugar and the sucrose is determined by the following formulæ, designating sucrose by the letter R, and invert-sugar by I.
$$R = \frac{100 \times \text{Polarization}}{\text{Polarization} + y}.$$
$$I = 100 - R.$$

Example.—Polarization of 26.048 grammes = 86.4. $p = 3.256$ grammes.

Suppose these 3.256 grammes have precipitated on boiling with Fehling's solution 0.290 grammes of copper. Then,

1. $\dfrac{Cu}{2} = \dfrac{0.290}{2} = 0.145 = Z.$

2. $\dfrac{Cu}{2} \times \dfrac{100}{p.} = 0.145 \times \dfrac{100}{3.256} = 4.45 = y.$

3. $\dfrac{100 \times Pol.}{Pol. + y} = \dfrac{8640}{86.4 + 4.45} = 95.1 = R.$

$100 - R = I,$

$100 - 95.1 = 4.9,$

$4.9 = I,$

and therefore the ratio of $R : I$ is expressed by 95.1 : 4.9.

In order to find the factor F we must hunt up the correct vertical and horizontal columns in Table XIII. The value $Z = 145$ is most closely approximated by the column headed 150; the ratio $R : I = 95.1 : 4.9$ is most closely approximated by the horizontal column 95 : 5. At the line of intersection of these two columns there will be found the factor 51.2, by aid of which the final calculation is effected.

4. $\dfrac{Cu}{p} \times F = \dfrac{0.290}{3.256} \times 51.2 = 4.56$ p. c. invert-sugar.

The analysis would hence show:

Sucrose, ... 86.40
Invert-sugar, ... 4.56

If duplicate or comparative determinations of invert-sugar are to be made by this method, the *same* weight of substance should always be taken. Otherwise, the value of Z varying, will necessitate the employing of different factors, and in consequence discrepancies will ensue.

Example :

Weight used, . . . 2.500 grammes.
Polarization, ... 95.00
Cu reduced, ... 0.140
Invert-sugar = 2.587 per cent.

Weight used, . . . 5.000 grammes.
Polarization, ... 95.00
Cu reduced, ... 0.278
Invert-sugar = 2.768 per cent.

Of the methods here described, Soxhlet's is possibly the most exact, but its execution calls for more time than can generally be given in a technical laboratory.

Of the other two methods given either may be used in practice, as each gives reliable results. Comparative determinations have shown that the results yielded by these two methods agree closely.*

If an invert-sugar determination has been made in a syrup, the result can be recorded either as percentage on the syrup, or as percentage on the dry substance. The calculation necessary to obtain the latter, corresponds of course, to that explained on page 41.

These methods of determining invert-sugar are based on the assumption that there are no other substances present besides invert-sugar which will precipitate the copper from its solution. Sometimes, however, such bodies are present. In beet-sugars their existence has been amply demonstrated, and their presence in cane-products is probable.

---

* The Author. "Determination of Invert-Sugar by Alkaline Copper Solutions," School of Mines Quarterly, November, 1888.

757

To determine the invert-sugar in such cases, a duplicate copper determination, the one before, the other after the destruction of the invert-sugar, is necessary.*

Of the caustic potash necessary for the preparation of Fehling's solution, dissolve 40 grammes, together with 175 grammes Rochelle salt, and make the solution up to 400 c.c. with water; 20 grammes of the caustic potash dissolve up separately with water to 100 c.c.

I. Heat 10 grammes (50 c.c.) of the sugar, clarified with basic acetate of lead, to boiling. Into this put 50 c.c. of Fehling's solution heated to the boiling-point. This solution is composed of 25 c.c. copper-sulphate solution, 20 c.c. of the alkaline Rochelle-salts solution, and 5 c.c. of the caustic-potash solution. Boil exactly two minutes.

II. 10 grammes (50 c.c.) of the sugar, clarified with basic acetate of lead, are boiled for 10 minutes with 5 c.c. of the caustic-potash solution, care being taken to replenish the water which evaporates. Then 25 c.c. copper-sulphate solution + 20 c.c. of the alkaline Rochelle-salts solution are added, and the mixture boiled for two minutes more. The rest of the determination is then carried out exactly as before described.

The amount of copper obtained under II. is subtracted from the amount found under I., and the remainder calculated to invert-sugar.

Soldaini's Solution.—Within the past few years great claims have been made for the Soldaini copper solution for the determination of invert-sugar, as being superior to the numerous so-called "Fehling" solutions. †

---

* Bodenbender and Scheller.

† Stammer's Jahresbericht, 1885, p. 283, enumerates no less than twenty different formulæ for the preparation of the same.

Soldaini's solution is prepared* by dissolving 15.8 grammes of sulphate of copper in a hot solution of 594 grammes of potassium bicarbonate. After the copper precipitate has completely dissolved, the solution is made up to 2 litres. The specific gravity of the solution is about 1.1789.

The manner of working with this solution is analogous to that described on page 69 *et seq.* The time of boiling is 10 minutes.

Table XIV shows the relation between the amount of copper reduced and the invert-sugar.

This solution has as yet not been generally adopted, but many opinions in its favor have been expressed.

Among the objections cited against it† are, that it contains only one fifth the amount of copper that Fehling's solution contains, and that hence it must be in many cases less sensitive than the former. On being greatly diluted it deposits cupric oxide, and on boiling for a long time it deposits cuprous oxide.

---

* Scheller's formula.

† Herzfeld, Zeitschrift des Vereines für Rübenzucker-Industrie, 1890, vol. xl. p. 52.

CHAPTER VI.

WATER—ASH—SUSPENDED IMPURITIES.

Water.—Weigh out 5 to 10 grammes of the sample. If the determination is to be made on a rather moist sugar or on a syrup, a corresponding amount of perfectly dry powdered glass or of sand must be intimately mixed with the sample.

Place in an air-bath, the heating of which should be commenced only after the introduction of the assay. The heat should be gradually carried up to between 95° and 100° C., and continued until the sample has attained to constant weight.

The loss in weight sustained, represents the water.

Example.—Weight of dish, sand, and sample, .   23.0000
    "   "   "   and sand,   . . .   18.0000

        Sample taken, ...   5.0000

Original weight of dish, sand, and sample, . .   23.0000
Final weight (after drying to constant weight),   21.1546

                 Water   =   1.8454

$$5.000 : 1.8454 :: 100 : x.$$

$x = 36.91$ per cent water.

Instead of drying in an air-bath, the drying can be done in a current of any inert gas, or it can be still more rapidly accomplished by drying in a vacuum. A tube provided with a number of small branch-tubes, each of which can be closed independently by means of a stop-cock, is put into connection with a vacuum-pump. The samples of sugar in which the moisture is to be determined, are weighed into metal dishes provided with a cover and of known weight, and these dishes, after being placed on a steaming water-bath, are connected with the branch-tubes and the air exhausted.

Entire dessication is accomplished in from half an hour to one hour's time.

A method for determining *approximately* the amount of water in a sample of syrup, liquor, or sweet-water, is to take the Brix hydrometer reading of the solution, and to subtract this from 100. The difference is accepted as representing the water.

Example.—Density of syrup in degrees Brix, 75°.0.

$$100$$
Less   75
$$\overline{25} \text{ per cent of water.}$$

Ash.—Scheibler's Method.—Weigh out 2.5 to 5 grammes of sample into a platinum ash-dish. Moisten with eight to ten drops of chemically pure conc. sulphuric acid, or better, with sixteen to twenty drops of dilute sulphuric acid (1 : 1). Pour a little ether over the contents of the dish and ignite. This treatment yields a porous carbonized mass, and avoids in a great measure the danger of loss by the assay mounting and creeping over the sides of the dish. When all gases have burned off, place in a platinum muffle, or in a Russia sheet-iron muffle (the metal should be about $\frac{1}{48}$ inch in thickness), and keep the muffle at a dull-red heat until the sample has been turned completely to ash; cool and weigh.

As the addition of sulphuric acid has converted a number of the salts present in the sugar into sulphates, 10 per cent is deducted from the weight of the ash found in order to make the results obtained by this method harmonize with those obtained by the method of carbonization.

Example.—Used 2.5 grammes of sugar.

Weight of dish + ash,	. .	13.9030
" " "	. .	13.8490

Ash, ...	. .	0.0540
Subtract 10 per cent,	. .	0.0054

Total ash,	=	0.0486

Total ash,	=	1.944 per cent.

This subtraction of one tenth of the weight of the ash is generally assumed to answer for beet-sugars, but is entirely misleading where cane-products are analyzed, because the ash of the latter possess a composition entirely different from the ash of the former.* At present, however, the subtraction of one tenth is still the general practice.

That unreliable results are obtained by this method of incineration with sulphuric acid and the subsequent subtraction of one tenth from the weight of the sulphated ash, even when beet-sugars are analyzed, has been recently admitted by European chemists of note.†

Von Lippmann ‡ advocates taking the dried-out sample, on which the water determination has been made, saturating it with vaselin-oil (having a boiling-point of about 400°),

---

* The Author, "Ash Determinations in Raw Sugars," School of Mines Quarterly, vol. xi. No. 1.

† Die Deutsche Zucker-Industrie, 1890, March 14, No. 11. Beilage 1, p. 007.

‡ Loc. cit.

and igniting the mixture. The carbonized mass is then to be burned to ash in a mixed current of air and oxygen. METHOD OF CARBONIZATION.—Weigh out 2.5 to 5.0 grammes of the sample. Carbonize at a low heat. Extract the soluble salts from the carbonaceous mass with boiling water; ignite the residue. Add the ash obtained to the aqueous extract and evaporate to dryness. Moisten with ammonium carbonate, drive off all ammonia, cool, weigh, and report as carbonate ash.

Quantitative Analysis of Sugar-Ash. — Dissolve 10 grammes of the sugar in hot water and filter;* wash the residue thoroughly with boiling water and evaporate the filtrate and the washings to dryness. Carefully carbonize the mass, and then extract the same with boiling water until nitrate of silver no longer gives the reaction for chlorine. Evaporate the solution to small bulk. The residue must be dried, ignited, and weighed. This weight is noted as, insoluble ash. The solution and the ash obtained are then combined, hydrochloric acid is added, and the solution evaporated to dryness. All the chlorine is then driven off, the residue is taken up with water and a little hydrochloric acid, and filtered. The insoluble residue in the filter is thoroughly washed, and the washings are added to the filtrate. This residue is silica. To the filtrate ammonic hydrate is added, and the solution is boiled and filtered; the residue, iron and alumina, must be thoroughly washed, and the washings added to the filtrate.

---

* This should be done in every case so as to have all the analyses made under the same conditions; in most instances it will be imperative, for the inorganic suspended impurities (sand, clay, etc.) in a sample of cane-sugar often weigh more than the total sugar-ash.

To this ammonium oxalate is added, and the whole is evaporated to dryness. The ammonia is burned off, and the oxalates are changed to carbonates by adding a little ammonium carbonate, and again driving off the ammonia.

The mass is then taken up with water, filtered, washed, and the washings added to the filtrate. The residue consists of the carbonates of calcium and magnesium. The filtrate is evaporated to small bulk, ammonium carbonate is added, and the evaporation is then continued to dryness, the ammonia is cautiously driven off, and the residue weighed. This gives the alkalies in the form of carbonates, and this weight added to the weight of the insoluble ash previously determined, represents the total carbonate ash.

Suspended Impurities.—It is often necessary to determine the share of work done in filtration respectively by the bag-filters and the bone-black.

The former, of course, remove only the mechanically suspended impurities, or at least the greater part of them, and leave to the bone-black the rest of the work to be accomplished.

The suspended impurities are both mineral and organic; their determination is effected in the following manner:

Dissolve from 2.5 to 10 grammes of the sample in hot water. Pour on a filter-paper which has previously been dried and weighed between watch-glasses, and wash with boiling water until all of the sugar has been removed. This is most conveniently done by the aid of a vacuum-pump. Then dry filter and contents to constant weight, and weigh as before between watch-glasses. The increase in weight over the previous weight, represents the total suspended impurities. Ignite the filter and contents in a platinum crucible, and record the weight of the ash as mineral or inorganic suspended impurities; the difference between the total suspended impurities and this figure gives the organic suspended impurities.

An ash determination made as previously described represents the mineral matter contained in the sugar, in the form of salts, etc., as well as the mineral matter mechanically suspended, and which latter, the bag-filters are supposed to remove.

The inorganic suspended impurities when subtracted from the total ash show the "soluble" ash, the more or less complete removal of which is expected of the bone-black.

Example.—Used 2.5 grammes of raw sugar.

Weight of watch-glasses + filter + total suspended impurities, ... ; 22.5071
Weight of watch-glasses + filter, ... 22.5000

Total suspended impurities, . 0.0071

Weight of crucible + ash of filter + inorganic suspended impurities, ... 13.20020
Weight of crucible, ... 13.20000

Ash of filter + inorganic susp. impurities, . 0.00020
Ash of filter, ... 0.00008

Inorganic susp. impurities, . 0.00012

Total suspended impurities, 0.00710 = 0.2840 per cent.
Inorganic " " 0.00012 = 0.0048 " "

Organic " " 0.00698 = 0.2792 " "

Total ash (previously determined), . 0.5040 per cent.

Inorganic suspended impurities, . . 0.0048 " "

Soluble ash, ... 0.4992 " "

Determination of Woody Fibre. —About 20 to 25 grammes of the sample, in as finely divided a state as possible, are placed in a flask or beaker, into which cold water is poured. The water, after having been in contact with the chips or shavings from 20 to 30 minutes, is decanted carefully, in order to avoid any loss of the weighed sample. This treatment with cold water is repeated two or three times, and is then followed by a similar treatment with hot water; finally, the sample is boiled several times, fresh water being taken for each treatment, and the treatment continued until all the soluble material has been washed out. Sometimes this is followed by washings with alcohol and ether.

. The sample is then transferred to a weighed filter, preferably made of asbestos, and gradually dried to constant weight. If dried in the air-bath, a temperature of 110° C. should not be exceeded. If the sample can be dried in vacuo, and subsequently weighed in a covered dish or capsule, all danger of oxidation and absorption of moisture are avoided.

The increase in weight which is noted in the filter, of course represents the woody fibre.

Detection of the Sugar-Mite.—To detect the sugar-mite (*Acarus sacchari*) in raw sugars, dissolve the sample in warm water; the mite will cling to the sides or to the bottom of the vessel. Drain off the solution and identify by means of a microscope.*

---

* For drawings, see Hassall, " Food and its Adulterations."

CHAPTER VII.

IN regular technical analyses the organic matter not sugar, raffinose, or invert-sugar is not determined. It is assumed to be represented by the difference between 100 and the constituents determined, viz., sucrose, raffinose, invert-sugar, water, and ash. This difference is frequently recorded as "non-ascertained," or "undetermined matter."

There are several methods for the direct determination of this organic matter, but the results which they yield are of value chiefly for comparative purposes. The following method is perhaps the most satisfactory:

Dissolve 10 to 20 grammes of raw sugar in warm water. Add basic acetate of lead solution in excess. Warm for a short time and filter. Wash the precipitate thoroughly; then suspend it in water and pass in sulphuretted hydrogen until all the lead is precipitated as sulphide. Filter out the sulphide of lead, wash thoroughly, and evaporate the filtrate and washings to dryness (constant weight), in a dish previously weighed. The temperature at which the drying is done, must not exceed 100° C.

Example.—Used 10 grammes of raw sugar.

Weight of dish and organic matter, ... 17.0973
"   " dish, ... 17.0482

Organic matter, ... 0.0491

Organic matter = 0.491 per cent.

83

The organic bodies accompanying sucrose can be divided into three classes:

1. Organic acids, or bodies that can act as acids.
2. Nitrogenous substances.
3. Non-nitrogenous substances.

These classes embrace respectively the following bodies:

ORGANIC ACIDS.*

Acetic, ...	$C_2H_4O_2$	Melassic, ...	$C_{12}H_{10}O_5$ (?)
Aconitic, ...	$C_6H_6O_6$	Metapectic, ..	$C_8H_{10}O_7$
Apoglucic, ..	$C_{18}H_{10}O_9$	Oxalic, ...	$C_2H_2O_4$
Aspartic, ...	$C_4H_7NO_4$	Oxycitric, ...	$C_6H_8O_8$
Butyric, ...	$C_4H_8O_2$	Parapectic,...	$C_{24}H_{30}O_{21}$
Citric, ...	$C_6H_8O_7$	Pectic, ...	$C_{16}H_{11}O_{13}$
Formic, ...	CH_2O_2	Propionic,...	$C_3H_6O_2$
Glucic, ...	$C_{12}H_{18}O_9$	Succinic, ...	$C_4H_6O_4$
Glutamic, ...	$C_5H_9NO_4$	Tartaric, ...	$C_4H_6O_6$
Lactic, ...	$C_3H_6O_3$	Tricarballylic, ..	$C_6H_8O_6$
Malic, ...	$C_4H_6O_5$	Ulmic, ...	$C_{24}H_{16}O_9$
Malonic, ...	$C_3H_4O_4$		

NITROGENOUS SUBSTANCES.

Albumin, ...	$C_{45}H_{72}N_{11}$ (?)	Legumin, ...	$C_{42}H_{64}N_{12}$(?)
Ammonia, ...	NH_3	Leucine, ...	$C_6H_{13}NO_2$
Asparagin,...	$C_4H_8N_2O_3$	Trimethylamin, .	C_3H_9N
Betaïne, ...	$C_5H_{11}NO_2$	Tyrosine, ...	$C_9H_{11}NO_3$
Glutamine,..	$C_5H_{10}N_2O_3$		

NON-NITROGENOUS SUBSTANCES.

Arabinose, ...	$C_8H_{12}O_8$	Pectin, ...	$C_{32}H_{48}O_{32}$
Cellulose, ...	$(C_6H_{10}O_5)_n$	Pectose, ...	$(C_2H_3O_2)_n$
Cholesterin, ..	$C_{26}H_{44}O$	Vanillin, ...	$C_8H_8O_3$
Coniferin, ...	$C_{16}H_{22}O_8$	Coloring matters,	
Dextrane, ...	$C_6H_{10}O_5$	Ethereal oils,	
Mannite, ...	$C_6H_8O_6$	Fats,	
Parapectin, ..	$C_{32}H_{48}O_{31}$	Gummy matters.	

* These acids are chiefly in combination with the metals potassium, sodium, calcium, magnesium, iron, and manganese. Rubidium and vanadium have also been identified in sugar-beets.

SCHEMES FOR ANALYSIS OF THE ORGANIC ACIDS.*

SCHEME   I. Non-volatile acids.

SCHEME II. Rare non-volatile acids.

SCHEME III. Volatile acids.

SCHEME IV. Approximate determination of organic acids, non-volatile and volatile.

---

* Translated by the author from the French of E. Laugier (Bittmann's arrangement), as published in Commerson and Laugier, Guide pour Analyse des Matières Sucrées, 3d Edition, 1884. Paris.

SCHEME I.

NON-VOLATILE ACIDS.

SCHEME I.
Non-Volatile Acids.

20 grammes of the sample are dissolved in distilled water. Add acetic acid in slight excess. Boil to complete expulsion of the carbonic acid, and neutralize with a dilute solution of ammonic hydrate free from carbonic acid. Precipitate with a slight excess of neutral acetate of lead; allow to digest for one hour. Bring the precipitate on a filter, and wash with boiling distilled water. The precipitate is then suspended in water and decomposed by sulphuretted hydrogen. The sulphide of lead is filtered out. The filtrate is boiled to expel the sulphuretted hydrogen. A few cubic centimetres of hydrochloric acid are added, and the solution filtered.

Precipitate.	Filtrate 1.
A white, gelatinous, transparent deposit, insoluble in alcohol: parapectin, pectin and parapectic acid. Brown flocks, soluble in alcohol: melassinic acid.	Make alkaline with ammonic hydrate; add ammonic chloride and calcic chloride. Filter out and wash the precipitate.

Precipitate.

May contain oxalic and tartaric acid, sulphate and phosphate of calcium. Treat in the cold with sodic hydrate. Filter and boil the filtrate. No precipitate: absence of tartaric acid. A precipitate: tartrate of calcium. Confirm by silver mirror, formed on addition of ammonic hydrate and argentic nitrate. That part of the precipitate which was not dissolved, is suspended in sufficient water to dissolve the calcium sulphate. Acidify with acetic acid: the phosphate of calcium dissolves, the oxalate of calcium does not. Oxalic acid is tested for by auric chloride, which is reduced, or by treatment with manganese dioxide and sulphuric acid, which yield carbonic acid.

Filtrate 2.

Three times the volume of 80-per-cent alcohol are added. No precipitate; absence of citric, malic, succinic, and glucic acids. If a precipitate is formed (the calcium salts of these acids), it is filtered out, washed with alcohol, dissolved in hydrochloric acid, and ammonic hydrate added to alkaline reaction. If on heating this solution no precipitate is formed, citric acid is absent; if a precipitate is formed, this is filtered out.

Precipitate.	Filtrate 3.
Precipitate. Citrate of calcium. It is less soluble in boiling than in cold water.	Three times the volume of 80-per-cent alcohol are added. A precipitate: succinate, malate, and glucate of calcium. Filter and divide the precipitate into two equal parts.

Precipitate: Part 1.

Dissolve in concentrated hydrochloric acid, and divide the solution into two parts.

Part 1.	Part 2.
Heat for a quarter of an hour to boiling. Brown flocks appear: Ulmic acid, formed from glucic acid. No brown flocks: absence of glucic acid.	Glucic acid is tested for by an alkaline solution of copper, which it reduces. Neutral and basic acetate of lead precipitate glucic acid incompletely in aqueous, but completely in alcoholic solution.

Precipitate: Part 2.

Dissolve by boiling in dilute nitric acid (1:1) and divide the solution into two parts.

Part 1.	Part 2.
Allow to cool. Colorless prismatic crystals of malate of calcium are deposited. Wash with water, dissolve by boiling, and precipitate by neutralization of lead acetate of calcium. Add calcic chloride. After twenty-four hours the amorphous lead salt turns crystalline and dissolves in hot water.	Evaporate with concentrated nitric acid, boil with a solution of carbonate of sodium, and filter out the precipitate. Neutralize the filtrate with nitric acid, boil several times, and add a solution of sulphate of calcium. The precipitate is oxalate of calcium. Add calcic chloride to precipitate completely the calcium. Oxalic acid derived from the malic acid, and in the filtrate test for succinic acid with ferric chloride.

SCHEME II.
RARE NON-VOLATILE ACIDS.

SCHEME II.
Rare Non-Volatile Acids.

Dissolve 20 grammes of the sample ; precipitate by neutral acetate of lead, place on a filter, and wash with boiled distilled water until the washings no longer contain lead.

Precipitate.	Filtrate 1.
It contains the lead salts of the organic acids, as well as the sulphate and phosphate of lead ; small quantities of parapectin may also be found in the lead precipitate. (Pectin is precipitated only by basic acetate of lead.) For the separation of these substances see column 2.	Add an excess of acetate of lead in solution, filter, and wash the precipitate.

Filtrate 1 splits into:

Precipitate.	Filtrate 2.
Suspend in water, pass sulphuretted hydrogen in excess, and filter out the sulphide of lead. From the filtrate remove the sulphuretted hydrogen by boiling, add alcohol and a few cubic centimetres of acetic acid. Filter.	This contains aspartic and metapectic acids. Add several cubic centimetres of an ammoniacal solution of acetate of lead, decompose by sulphuretted hydrogen, and filter out the sulphide of lead. Evaporate the filtrate to small bulk; add an equal volume of nitric acid (sp gr. 1.42), and heat for a quarter of an hour. Aspartic acid remains unchanged ; metapectic acid is decomposed into oxalic acid, which goes into solution, and into mucic acid, which crystallizes on cooling. Filter.

The precipitate splits into:

Precipitate.	Filtrate.
Pectin and parapectin. These substances may be separated in the same manner as legumin. To effect this, acidify strongly with acetic acid, boil, and filter out the coagulum.	This may contain small quantities of glucic, malic, and succinic acids which were not completely precipitated by neutral acetate of lead. Besides these there may be present traces of aspartic and of metapectic acids, which may be identified after the precipitation of the former acids, by nitrate of calcium and alcohol. (See the following column.)

Filtrate 2 splits into:

Crystals.	Mother Liquor.
The washed crystals of mucic acid are boiled with nitric acid; the mucic acid is decomposed completely into oxalic and tartaric acids, the identification of which proves the presence originally of mucic acid.	This contains aspartic and oxalic acids produced by the foregoing decomposition. Pass a current of N_2O_3. Nitrogen is set free, and at the same time malic acid is formed (at the expense of the aspartic acid). This is searched for as directed in Scheme I. The identification of malic acid proves the existence of aspartic acid in the original solution.

SCHEME III.
VOLATILE ACIDS.

SCHEME III.

Volatile Acids.

20 to 100 grammes of the sample (syrups, etc., are brought to 20° Baumé) are rendered strongly acid by dilute sulphuric acid. All the chlorine of the metallic chlorides is precipitated with a standardized sulphate of silver solution, and the precipitate of argentic chloride is filtered out. The liquid is distilled as long as acid vapors pass over, the distillate is exactly saturated with a solution of barium hydrate, and any excess of this reagent which might have been added, is removed by a stream of carbonic-acid gas. The liquid is concentrated, the barium carbonate filtered out, and the filtrate evaporated to dryness at 110° C. in a platinum capsule.

Residue of Distillation.	Distillate.			
Contains nearly the whole of lactic acid, only traces having passed over into the distillate. Add three volumes of alcohol and distil the mixture with milk of lime. Filter the boiling solution to separate the hydrate and sulphate of calcium. In this filtrate the lime is precipitated by a stream of carbonic-acid gas. Evaporate to dryness, take up the residue with strong alcohol, filter again, and let the filtrate stand. If lactic acid is present, crystals of calcium lactate are formed, which are recognized by their characteristic structure.	The dried barium salts obtained from the distillate are extracted with boiling alcohol of 84 per cent, the operation being repeated several times, and the residue remaining undissolved, is filtered out.			
	Residue.		**Solution.**	
	Formate and nitrate of barium. Traces of acetate of barium. Dissolve in a little water, and precipitate the barium with sulphate of sodium. Filter, and mix a portion of the filtrate with argentic nitrate. Citrate of silver, which is precipitated, is reduced by heating to a mirror of metallic silver. In another portion of the solution test for formic acid by the reduction of mercuric to mercurous chloride.		Acetate, propionate, and butyrate of barium. Evaporate to small bulk, take up with a little water, precipitate the barium with sulphuric acid, filter out the precipitate, and divide the filtrate into two equal parts. Neutralize one portion with sodium hydrate, and then add this to the other portion. Subject the whole to distillation.	
			Distillate.	**Residue.**
			Butyric and propionic acids. They are identified by their odor, and the oily drops which are formed in decomposing their salts by sulphuric acid.	Acetic acid. Identified by its odor, and by the formation of acetic ether, produced on warming one of its salts with sulphuric acid and alcohol.

SCHEME IV.

APPROXIMATE DETERMINATION OF ORGANIC ACIDS: NON-VOLATILE AND VOLATILE.

SCHEME IV.

Approximate Determination of Organic Acids, Non-Volatile and Volatile.

Non-volatile Acids.			Volatile Acids.
A. Precipitation by neutral acetate of lead. Oxalic, citric, tartaric, and malic acids. Incompletely: pectic, parapectic, glucic, melassinic, ulmic, and succinic acids.	B. Precipitation by basic acetate of lead. Pectic, parapectic, glucic, melassinic, ulmic, and succinic acids. Parapectin. In-	C. Precipitation by ammoniacal acetate of lead. Aspartic and metapectic acids.	D. Not precipitated by acetate of lead: formic, acetic, lactic, propionic, and butyric acids.
50 grammes of the sample are dissolved in distilled water and made slightly acid with acetic acid. The solution is boiled to expel the carbonic acid, and neutralized with sodium hydrate (free from carbonic acid). A slight excess of neutral acetate of lead is added, and digested for one hour. The residue is placed on a dry and weighed filter, and is washed with boiled distilled water until the washings give no longer the reaction for lead. (For treatment of the filtrate, see B.) The precipitate contains the lead salts of the above-named acids, and besides sulphate and phosphate of lead, if the sample examined contained sulphates and phosphates. The filter with its contents is dried at 110° C., and weighed. The precipitate is removed, the filter is burned in a weighed platinum crucible, the precipitate is again added, and heated to dull redness. To facilitate the combustion of the carbon, small doses of ammonium nitrate are repeatedly added, great care being taken to prevent loss by spitting. After cooling, the crucible is weighed. The weight of the contents of the crucible subtracted from that of the precipitate dried at 110° C. represents the weight of the organic acids, because the sulphate and phosphate of lead are not altered by the ignition.	completely: aspartic and metapectic acids, and pectin. — — To the filtrate from the lead salts precipitated by neutral acetate of lead, there is added a slight excess of basic acetate of lead, and the precipitate filtered out. (For filtrate, see C.) The precipitate is placed on a dried and weighed filter, then washed, dried at 110° C., and weighed. A part is incinerated as in A, and the weight of the organic acids determined by difference, as there described.	— The filtrate obtained from the precipitation with basic acetate of lead is mixed with several cubic centimetres of an ammoniacal acetate of lead solution. Allow to stand for twelve hours. Filter, allow to drain off, and wash once with distilled water to which a little ammoniacal acetate of lead has been added. The precipitate, dried and weighed, is treated as described under A and B. Note. — The ammoniacal acetate of lead must be added only gradually and in small amounts, for without this precaution it is apt to precipitate sugar, and then even an approximate determination of the acids sought for, becomes very difficult.	— 50 grammes of the sample to be examined (in case of juices a larger amount must be taken; thick syrup must be diluted), are strongly acidified with dilute sulphuric acid. All the chlorine which has been previously determined volumetrically in a separate sample, is precipitated by a standardized sulphate of silver solution. The filtrate from the argentic chloride is distilled until acid fumes no longer pass over. This distillate is then mixed with a solution of barium hydrate, any excess of this reagent is precipitated by carbonic acid, and the solution filtered. The filtrate is evaporated to dryness at 110° C. in a weighed platinum capsule; the residue represents the weight of the organic acid salts of barium, which are determined as sulphates or carbonates. If nitrates were present in the sample analyzed, the residue contains also barium nitrate. In that case the nitric acid must be determined, the weight of the barium nitrate calculated from the result, and this value subtracted from the weight of the organic acid salts of barium previously found.

Determination of Total Nitrogen.*—An amount of the substance, varying from 0.7 to 2.8 grammes, according to its proportion of nitrogen, is placed in a digestion-flask with approximately 0.7 gramme of mercuric oxide and 20 cubic centimetres of sulphuric acid. †

The flask is placed in an inclined position, and heated below the boiling-point of the acid, from five to fifteen minutes, or until frothing has ceased. The heat is then raised until the acid boils briskly, and this boiling is continued until the contents of the flask have become a clear liquid, colorless, or of a very pale straw color.

While still hot, finely pulverized potassium permanganate is introduced carefully and in small quantity at a time, till, after shaking, the liquid remains of a green or purple color.

After cooling, the contents of the flask are transferred to the distilling-flask, with about 200 cubic centimetres of water; to this a few pieces of granulated zinc and 25 cubic centimetres of potassium-sulphide solution ‡ are added, shaking the flask to mix its contents. Sufficient of a sodium hydrate solution§ is then added to make the reaction strongly alkaline. This reagent should be poured down the sides of the flask, so that it does not mix at once with the acid solution.

The flask is then connected with the condenser, and its contents are distilled until all ammonia has passed

---

* The Kjeldahl method. Abstracted from Bulletin No. 19, U. S. Department of Agriculture.

† C. P. acid, specific gravity 1.83, free from nitrates and ammonium sulphate.

‡ Prepared by dissolving 40 grammes of commercial potassium sulphide in 1 litre of water.

§ A saturated solution of sodium hydrate, free from nitrates.

over into standard hydrochloric acid.* The distillate is then titrated with standard ammonia.

Previous to use, the reagents should be tested by a blank experiment with sugar, which will partially reduce any nitrates that are present, and which might otherwise escape notice.

If the nitrogen present in *organic* combination is to be ascertained, the nitrogen present in the form of nitric acid and in the form of ammonia must be separately determined, and their sum subtracted from the total nitrogen found; the remainder is the nitrogen in organic combination.

Non-Nitrogenous Organic Substances.—The determination of non-nitrogenous organic substances is effected by aid of basic and neutral acetate of lead and alcohol (pectin and parapectin), by the successive use of water, alkalies, acids, alcohol, and ether (cellulose), by treatment with ether (fats, essential oils), by the aid of yeast fermentation, and alcohol (isolation of mannite).†

Determination of Pure Cellulose.‡—To make this determination, place 10 grammes of the sample, 30 to 40 grammes of pure potassium hydrate, and about 30 to 40 c.c. of water into a glass retort. Close the retort by a glass stopper, place in an oil-bath, provided with a thermometer, and heat up gradually. At about 140° C. the solution will commence to boil and foam considerably. Increase the temperature to about 180°, and continue heating for about one hour. When the contents of the

---

* Half-normal acid, 18.25 grammes HCl to the litre.

† For details of these determinations see Zeitschrift des Vereines für Rübenzucker-Industrie, 1879, vol. xxix. p. 906.

‡ Method of G. Lange. Chemisches Repertorium, 1890, vol. xiv., No. 3, p. 30.

retort cease foaming, become quiet, and begin to turn dry, the end of the reaction has been reached.

Remove the retort from the oil-bath, and after cooling to about 80°, add hot water and rinse the contents of the retort carefully first with hot and then with cold water, into a beaker.

After cooling, acidify with dilute sulphuric acid ; this acid will precipitate the particles of cellulose which have been kept in suspension in the strong alkaline solution. Then, with very dilute sodium hydrate, produce anew a faintly alkaline reaction, so that all of the precipitated substances, excepting the cellulose, may be again brought into solution.

The residue is then transferred to a weighed filtering tube provided with a finely perforated platinum cone and washed out thoroughly, first with hot water, and then with cold. Drying is effected on a water-bath, and the filter with its contents weighed.

The residue is then removed from the filter, ignited, and the weight of the ash found subtracted from the value previously obtained. The difference in weight represents pure cellulose.

CHAPTER VIII.

NOTES ON THE REPORTING OF SUGAR-ANALYSES, DETERMINATION AND CALCULATION OF THE RENDEMENT, ETC.

In commercial analyses it is customary to report only—

Polarization,
Invert-sugar,
Water,
Ash,
Non-ascertained,

the "non-ascertained" being the balance required to make the analysis figure up to 100.

When beet-sugars are examined, and a raffinose determination has been made, this substance, of course, makes another item in the report, which would then embrace:

Polarization,
Sucrose,
Raffinose,
Invert-sugar,
Water,
Ash,
Non-ascertained.

The polarization in the first form of analysis given above, may either correspond to, be greater, or smaller than the amount of sucrose really present, for the presence of other optically-active bodies influences the polariscope-reading to a marked degree.

Invert-sugar turns the plane of polarized light to the left. At 17°.5 C. one part of invert-sugar neutralizes the optical effect of 0.34 parts of sucrose. In order, therefore, to obtain the sucrose corrected for this disturbing influence, the amount of invert-sugar found is multiplied by 0.34, and the result is added to the direct polarization. This sum is then regarded as representing the sucrose.

Frequently a polarization after inversion is made, and compared with the direct polarization.

If there are no other optically active bodies present in the sample besides the sucrose, the result of the polarizations before and after inversion will be identical, or at least agree very closely. If the polarization after inversion is higher than the direct polarization, the presence of lævo-rotary bodies is indicated; if it is lower, dextro-rotatory substances are present.

Recent investigations have, however, shown that this method of inversion and subsequent polarization (Clerget's test) is not applicable to sugars rich in reducing sugars (so-called invert-sugar), because the inverting acid (hydrochloric acid) increases the lævo-rotation of the invert-sugar,* and because the reducing sugar sometimes consists of a mixture of lævo- and of dextro-rotatory substances in varying proportions.

In dealing with samples of such description, as, for instance, low sugars and molasses, sugar-cane products, an exhaustive analysis is desirable, in order to gain all information possible with regard to the nature of the sample.

---

* Jungfleisch and Grimbert, Report to the French Academy of Sciences, December, 1889.

Such an analysis should record—

Reaction (acid, alkaline, or neutral),
Total sucrose,
Polarization after inversion,
Direct polarization,
Total reducing sugars,
Water,
Ash.

The interpretation of an analysis of this description is not always an easy matter.

If the polarization after inversion agrees with the direct polarization *plus* 0.34 times the total reducing sugar, this value *may* be regarded as the amount of sucrose (crystallizable sugar) present. As, however, all results obtained by the Clerget method on sugars rich in invert-sugar are open to doubt, it will be better, even in case the direct polarization *plus* 0.34 times the total reducing sugar is equal to the polarization after inversion, to resort to gravimetric determinations for verification of the result.

In case of non-agreement of the direct polarization *plus* 0.34 times the total reducing sugar, and the Clerget test, of course gravimetric analysis must be employed.

Determine the total sucrose, after inversion, by its reducing action on copper solution, and in a similar manner determine also the total reducing sugar. Calculate the latter over to its equivalent of sucrose by subtracting one twentieth of the amount found; deduct this result from the total sucrose, and report the remainder as sucrose.

Example.—

Polarization before inversion, . . . 52.70

Polarization after inversion, . . . 63.12

Total reducing sugar, ... 22.89

Total sucrose (gravimetric det.), . . 79.20

	22.89	Total sucrose,	79.20
Less $\frac{1}{20}$, .	1.14	Less . .	21.75
	21.75	Sucrose =	57.45

Concerning the nature of the reducing sugar, this may be present as—

a. Optically Inactive Sugar.—The existence of a sugar that will reduce copper solution, but which is inactive to polarized light, is, at best, doubtful. But it might happen that the lævo-rotatory power of the invert-sugar is just neutralized by the dextro-rotatory influence of some other substance—raffinose or dextrose, for instance.* In either case the direct polarization and the polarization after inversion would agree.

b. Invert-Sugar.—In this case, barring the danger of an increased lævo-rotation by the inverting acid, the polarization after inversion will be equal to the sum of the direct polarization *plus* 0.34 times the reducing sugar.

c. Dextrose (Glucose).—In this case the polarization after inversion is equal to the direct polarization *minus* the reducing sugar multiplied by a factor. This factor has been given as 0.8. This seems, however, to be correct only when the dextrose, which is a bi-rotatory substance, has reached its lowest rotatory value, for experiments made by the author on mixtures of anhydrous crystallized dextrose and raw sugars of various grades,

---

* Bornträger, Deutsche Zuckerindustrie 1890, p. 277, claims, that owing to bi-rotation of the dextrose of the anhydrous invert-sugar, the lævo-rotation of the lævulose is temporarily neutralized.

gave values that fluctuated considerably from the factor quoted.

d. Mixture of Invert-Sugar and Dextrose, or Invert-Sugar and Lævulose, in varying proportions:

In this case only an analysis of the reducing sugar (see page 61) will permit a conclusion as to its composition. In all cases a gravimetric determination of the invert-sugar, the dextrose, or lævulose will afford a valuable check on any inferences that may be drawn from the data obtained by optical analysis.

If a cane-juice has been analyzed, the report should embrace the following determinations: *

1. Density expressed as specific gravity, or in degrees of Baumé or Brix.

2. Total solids.

3. Sucrose.

4. Reducing sugar (glucose).

5. Solids not sugar.

6. Coefficient of purity.

7. Glucose ratio.

No. 5 is equal to No. 2, less No. 3 + No. 4.

No. 6 is found by multiplying No. 3 by 100, and dividing by No. 2.

No. 7 is obtained by multiplying No. 4 by 100, and dividing by No. 3.

The percentage of extraction is obtained by dividing the weight of juice obtained by weight of cane used, and multiplying by 100.

Rendement.—The yield in crystallizable sugar can be analytically determined by the Payen-Scheibler method.

This process is based on the treatment of the raw

---

* Scheme adopted by the Louisiana Sugar Association.

sugar, whose rendement is to be ascertained, by solutions that will wash out the molasses-forming impurities, and leave behind the pure crystallizable sugar.

Five solutions are required:

No. 1 is a mixture in equal parts, by volume, of absolute alcohol and ether.

No. 2 is absolute alcohol.

No. 3 is alcohol of 96 per cent Tralles.*

No. 4 is alcohol of 92 per cent Tralles.

No. 5 is alcohol of 85 per cent to 86 per cent Tralles, to which 50 c.c. of acetic acid per litre have been added.

Solutions Nos. 3, 4, and 5 are all *saturated* with pure sugar; and, in order that they should remain saturated with sugar at all temperatures, they are kept in flasks which are half filled with best granulated sugar, previously washed with absolute alcohol.

These flasks are provided with a siphon arrangement; the air enters through chloride-of-calcium tubes, so as to be thoroughly dried; the solution is discharged through tubes filled with pure and dry sugar. Plugs of felt placed at the ends of these tubes prevent the carrying over of any sugar particles.

The washing operation is carried out as follows: The accurately weighed sample, usually 13.024 grammes, is placed into a 50 c.c. flask which has previously been dried.

A cork or a rubber stopper, through which two glass tubes are made to pass, serves to close the flask. One of these tubes reaches down almost to the bottom of the flask; it is provided with a felt-plug at its mouth; this

---

* The alcoholometer of Tralles gives the percentage volume for the temperature of 60° F. = 15⅝° C. Watt's Dictionary of Chemistry, vol. i. p. 84.

serves as strainer. The shorter tube only reaches to just below the cork or stopper. The longer tube is connected, by means of a rubber tube, with a large receiving bottle, from which the air is to a great extent exhausted by an aspirator or a vacuum pump. The rubber tube is provided with a pinch-cock, so that connection can be made or broken at will, between the receiving bottle and the small flask which holds the sample.

The apparatus being thus arranged, about 30 c.c. of solution No. 1 is allowed to flow into the flask containing the sugar. This solution is permitted to remain quietly in contact with the sample for from fifteen to twenty minutes, and is then drawn over into the receiving bottle. When it has all been drained over, 30 c.c. of solution No. 2 are introduced. After a contact of two minutes this solution is drawn off, and followed successively by about the same amounts of the other three solutions, in the order of their numbering.

The last of these, solution No. 5, is really the active reagent, the others principally serving to displace the moisture contained in the sugar.

This solution is allowed to remain on the sample for half an hour, being frequently and well shaken in the mean time to insure intimate contact.

It is then drawn off, and replaced by a fresh supply of the same solution. This in turn is drawn off, and the treatment is repeated with fresh amounts of solution No. 5, until the solution standing above the sugar, remains perfectly colorless. The time of contact is thirty minutes for each treatment.

The last traces of the solution No. 5 are then removed by successive addition of solutions Nos. 4, 3, and 2, in the order named. These are added and drawn off at intervals of two minutes each. The last traces of alcohol are removed by drying on a water-bath, a current of dry air being continuously drawn through the flask in the mean time. When the sample is perfectly dry, the cork with its inserted tubes is carefully withdrawn, and any sugar clinging to the long tube or its felt plug, is carefully washed into the flask. The solution is then made up to 50 c.c. and polarized. The reading on the polariscope represents in percentage the yield in crystallizable sugar.

Calculation of Rendement. — United States of America.—From the polarization (the crystallizable) subtract five times the ash, for sugars of all grades.

If the sugars are products of the beet, then, in addition to the above, subtract for—

1st Products: Three times the invert-sugar (non-crystallizable), if it does not exceed one quarter per cent; five times the invert-sugar (non-crystallizable), if it exceeds one quarter per cent.

2d Products: Three times the invert-sugar (non-crystallizable), if it does not exceed one half per cent; five times the invert-sugar (non-crystallizable), if it exceeds one half per cent.

England.*—Beet-Sugars.—1st. Products. *Basis,* 88 *p. c.*—From the crystallizable sugar deduct five times the ash and three times the non-crystallizable, provided the latter does not exceed one quarter per cent. If it exceeds this amount, then subtract five times the non-crystallizable.

Lower Products. *Basis,* 75 *p. c.*—From the crystal-

---

* Liste Générale des Fabriques de Sucre. Paris, 1889.

lizable, deduct five times the ash and three times the non-crystallizable, provided it does not exceed one half per cent. If it exceeds this limit, deduct five times the non-crystallizable.

FRANCE.* —Beet-Sugars. —From the crystallizable sugar subtract four times the ash and twice the non-crystallizable, which must not exceed one quarter per cent. From this rendement, figured without fractions of a degree, subtract one and one half per cent.

GERMANY.—From the crystallizable sugar (as determined by the polariscope), subtract five times the salts, i.e., the ash less the suspended impurities, and twice the invert-sugar.

Duty.—The duty levied by the United States Government is based on the polariscope test and on color.

For the color-test the "Dutch standards" (see page 25) have been adopted as the guide. In testing by the polariscope every fraction over a full degree is figured as if the next whole degree had been indicated. Thus, a sugar testing 94.0 degrees on the polariscope pays the duty prescribed for this grade, but a sugar testing 94.1 is classed as a 95.0 sugar.

The following is quoted from the existing law (March, 1890):

"All sugars not above No. 13 Dutch standard in color, . . . testing by the polariscope not above 75°, shall pay a duty of $1\frac{4}{10}$ cent per pound, and for every additional degree, or fraction of a degree, shown by the polariscope test, they shall pay $\frac{4}{100}$ of a cent per pound additional.

"All sugars above No. 13 Dutch standard shall be classified by the Dutch standard of color, and shall pay duty as follows, namely : All sugar above No. 13 and not above No. 16, 2¾ cents per pound ; all above No. 16 and not above No. 20, 3 cents ; all above No. 20, 3½ cents."

Calculation of the Weight of Solids and Liquids from their Specific Gravity.—One cubic foot of distilled water weighs 62.50 lbs. = 1000 ounces. The specific gravity of water is 1.000. If the decimal point of a specific-gravity value be moved three places to the right, the weight of a cubic foot in ounces will be obtained. This value divided by 16 gives the weight of a cubic foot in pounds. From this the following rule is deduced :

To find the weight in pounds per cubic foot :

Determine the specific gravity. Remove the decimal point three places to the right, and divide by 16.

Example.—Specific gravity of a bone-black is 0.87904.

$$879.04 \div 16 = 54.94.$$

Hence the bone-black weighs 54.94 lbs. per cubic foot.

As above stated, if the decimal point of a specific-gravity value is removed three places to the right, the weight of a cubic foot in ounces will be obtained, and this figure divided by 16 will give the weight of a cubic foot in pounds. But if the cubic foot be assumed equal to 7.5 gallons, $7.5 \times 16 = 120$. Therefore,

To find the weight of a gallon in pounds :

Determine the specific gravity. Remove the decimal point three places to the right, and divide by 120.

Example.—A syrup has a specific gravity of 1.413.

$$1413 \div 120 = 11.78.$$

Hence the syrup weighs 11.78 lbs. per gallon.

CHAPTER IX.

SYNONYMS—LITERATURE ON SUGAR ANALYSIS—TABLES.

SYNONYMS.

English.	German.	French.
Cane-sugar Saccharose Sucrose Common sugar Crystallizable sugar Diglucosic alcohol	Rohrzucker Saccharose Sucrose Saccharobiose	Sucre de Canne Saccharose Sucrose Sucre-normal Sucre Saccharon Cannose
Dextrose Glucose Glycose Fruit sugar Honey sugar Diabetic sugar Uric sugar Rag sugar Potato-sugar Right-handed sugar Grape sugar Starch sugar Dextro-glucose Sucro-glucose	Dextrose Glucose Glycose Honigzucker Harnzucker Traubenzucker Stärkezucker Krümelzucker	 Glycose Sucre de Raisin
Levulose (lævulose) Fruit sugar Left-handed glucose Lævo-glucose Sucro-glucose	Lävulose Fruchtzucker Linksfruchtzucker Syrupzucker Schleimzucker Honigzucker Chylariose	Lévulose Chyliarose

English.	*German.*	*French.*
Invert-sugar	Invertzucker	Sucre ínverti Sucre interverti
Raffinose Melitose Plus-sugar	Raffinose Melitose Melitriose Pluszucker Gossypose Baumwollzucker Raffinotriose Raffinohexose	Raffinose Melitose

REFERENCES TO LITERATURE

ON

SUGAR ANALYSIS.

BOOKS AND PERIODICALS.

1839 PELIGOT, E. Analyse et Composition de la Betterave à Sucre.

1840 PELIGOT, E. Composition chimique de la Canne à Sucre.

1848 *BACHE, A. D., AND McCULLOUGH, R. S. Report on Sugar and Hydrometers.

1863 FRESE, O. Beiträge zur Zuckerfabrikation.

1865 ICERY, E. Recherches sur les Jus de la Canne à Sucre.

1867 *MANDELBLÜH, C. Leitfaden zur Untersuchung der verschiedenen Zuckerarten, sowie der in der Zuckerfabrikation vorkommenden Produkte.

1867 MONIER, E. Guide pour l'Essai et l'Analyse des Sucres.

1868 *VIOLETTE, C. Dosage du Sucre au Moyen des Liqueurs titrées.

1869 MOIGNO, L'ABBÉ. Saccharometrie optique, chimique et melassimetrique.

1874 POSSOZ, L. Notice sur la Saccharometrie chimique.

1875 GUNNING, J. W. La Saccharometrie et l'Impot sur le Sucre.

1875 TERREIL, M. A. Notions pratiques sur l'Analyse chimique des Substances saccharifères.

1875 WACKENRODER, B. Anleitung zur chemischen Untersuchung technischer Produkte welche auf dem Gebiete der Zuckerfabrikation und Landwirthschaft vorkommen.

1876 MAUMENÉ, E. J. Traité théorique et pratique de la Fabrication du Sucre.

1878 *URE'S Dictionary of Arts, Manufactures, and Mines, vol. iii., and Supplement (1879).

Asterisks mark the publications consulted.—F. G. W.

1879 BARBET, E. Analyse des Liquides Sucrés.

1879 *LANDOLT, H. Das optische Drehungsvermögen Organischer Substanzen und die praktischen Anwendungen desselben.

1880 COLLIER, P. Report of Analytical and Other Work done on Sorghum and Cornstalks. Department of Agriculture, Report No. 33.

1881 FRANKEL, J., AND HUTTER, R. A. Practical Treatise on the Manufacture of Starch, Glucose, Starch-sugar, and Dextrine.

1882 *LANDOLT, H. Handbook of the Polariscope and its Practical Applications. (From the German.)

1882 *VON LIPPMANN, E. Die Zuckerarten und ihre Derivate.

1882 *SPONS' Encyclopædia of the Industrial Arts, Manufactures, and Raw Commercial Products, vol. ii., article: "Sugar Analysis."

1883 LE DOCTE, A. Traité complet du Contrôle chimique de la Fabrication du Sucre.

1883 LEPLAY, H. Chimie théorique et pratique des Industries du Sucre.

1883 *TUCKER, J. H. A Manual of Sugar Analysis. (Second Edition.)

1884 *COMMERSON, E., ET LAUGIER, E. Guide pour Analyse des Matières sucrées. (Third Edition.)

1884 *VON WACHTEL, A. Hilfsbuch für chemisch-technische Untersuchungen auf dem Gesammtgebiete der Zuckerfabrikation.

1885 *ALLEN, A. H. Commercial Organic Analysis, vol. i., article: "Sugars."

1885 *FRÜHLING, R., UND SCHULZ, J. Anleitung zur Untersuchung der für die Zuckerindustrie in Betracht kommenden Rohmaterialien, Producte, Nebenproducte und Hülfssubstanzen. (Third Edition.)

1887 *Ausführungs-Bestimmungen zum Zucker-steuergesetz vom 9ten Juli, 1887. (German Government.)

1887 *SCHMIDT, F., UND HAENSCH. Gebrauchs-Anweisung zu den Polarisations-Apparaten von Schmidt und Haensch.

1887 *STAMMER, K. Lehrbuch der Zuckerfabrikation. (Second Edition.)

1888 LOCK AND NEWLAND. Sugar: A Handbook for Planters and Refiners.

1888 PELLET. Nouveau Procédé simple, rapide et peu coûteux de Dosage direct du Sucre contenue dans la Betterave, la Canne, la Bagasse, le Sorgho, etc.
1888 *SACHS, F. Revue Universelle des Progrès de la Fabrication du Sucre.
1888 *TOLLENS, B. Kurzes Handbuch der Kohlen-hydrate.
1888 *WEIN, E. Tabellen zur quantitativen Bestimmung der Zuckerarten.
1889 *BASSET, N. Guide du Planteur de Cannes.
1889 *LEPLAY, H. Études chimiques sur la Formation du Sucre.
1889 *SPENCER, G. L. A Handbook for Sugar Manufacturers and their Chemists.

PERIODICALS.

*The American Chemist (1870-1877).
*The Louisiana Planter and Sugar Manufacturer. America. Weekly.
Sugar Bowl and Farm Journal. America. Weekly.
The Sugar Beet. America. Monthly.
*Sugar Cane. England. Monthly.
Sugar. England. Monthly.
The Journal of the Society of Chemical Industry. England. Monthly.
*Chemiker Zeitung. Semi-weekly.
*Die Deutsche Zuckerindustrie. Weekly.
*Jahresbericht über die Untersuchungen und Fortschritte auf dem Gesammtgebiete der Zuckerfabrikation.
*Neue Zeitschrift für Rübenzucker-Industrie. Semi-monthly.
*Oesterreichisch-Ungarische Zeitschrift für Zucker-Industrie und Landwirthschaft. Six numbers per annum.
Taschenkalender für Zuckerfabrikanten. K. Stammer. Annual.
Wochenschrift des Centralvereines für Rübenzucker-Industrie in der Oester: Ungar: Monarchie.
*Zeitschrift des Vereines für die Rübenzucker-Industrie des Deutschen Reichs. Monthly.
Zeitschrift für Zuckerindustrie in Böhmen. Ten numbers per annum.
Bulletin de l'Association Belge des Chimistes. Monthly.
*Journal des Fabricants de Sucre. France. Weekly.
*La Sucrerie Indigène et Coloniale. France. Weekly.

TABLES.

I.

RELATION BETWEEN SPECIFIC GRAVITY, DEGREES BRIX AND DEGREES BAUMÉ, FOR PURE SUGAR SOLUTIONS FROM 0 TO 100 PER CENT.

(Temperature 17.5° C. = 63.5° F.)

MATEGCZEK AND SCHEIBLER.

$$\text{Sp. Gr.} = \frac{100}{100 - (0.6813 \times \text{Degree Baumé})}.$$

$$\text{Degree Baumé} = 1.46778 \times \text{F.*}$$

$$\text{Degree Brix} = 259.3 - \frac{259\,3}{\text{Specific Gravity}}.$$

* The values of F. are given in Zeitschrift des Vereines für Rübenzucker-Industrie, 1865, page 580; 1870, page 263; 1874, pages 843 and 950.

I.

Degrees Brix.	Specific Gravity.	Degrees Baumé.	Degrees Brix.	Specific Gravity.	Degrees Baumé.
0.0	1.00000	0.00	4.0	1.01570	2.27
0.1	1.00038	0.06	4.1	1.01610	2.33
0.2	1.00077	0.11	4.2	1.01650	2.38
0.3	1.00116	0.17	4.3	1.01690	2.44
0.4	1.00155	0.23	4.4	1.01730	2.50
0.5	1.00193	0.28	4.5	1.01770	2.55
0.6	1.00232	0.34	4.6	1.01810	2.61
0.7	1.00271	0.40	4.7	1.01850	2.67
0.8	1.00310	0.45	4.8	1.01890	2.72
0.9	1.00349	0.51	4.9	1.01930	2.78
1.0	1.00388	0.57	5.0	1.01970	2.84
1.1	1.00427	0.63	5.1	1.02010	2.89
1.2	1.00466	0.68	5.2	1.02051	2.95
1.3	1.00505	0.74	5.3	1.02091	3.01
1.4	1.00544	0.80	5.4	1.02131	3.06
1.5	1.00583	0.85	5.5	1.02171	3.12
1.6	1.00622	0.91	5.6	1.02211	3.18
1.7	1.00662	0.97	5.7	1.02252	3.23
1.8	1.00701	1.02	5.8	1.02292	3.29
1.9	1.00740	1.08	5.9	1.02333	3.35
2.0	1.00779	1.14	6.0	1.02373	3.40
2.1	1.00818	1.19	6.1	1.02413	3.46
2.2	1.00858	1.25	6.2	1.02454	3.52
2.3	1.00897	1.31	6.3	1.02494	3.57
2.4	1.00936	1.36	6.4	1.02535	3.63
2.5	1.00976	1.42	6.5	1.02575	3.69
2.6	1.01015	1.48	6.6	1.02616	3.74
2.7	1.01055	1.53	6.7	1.02657	3.80
2.8	1.01094	1.59	6.8	1.02697	3.86
2.9	1.01134	1.65	6.9	1.02738	3.91
3.0	1.01173	1.70	7.0	1.02779	3.97
3.1	1.01213	1.76	7.1	1.02819	4.03
3.2	1.01252	1.82	7.2	1.02860	4.08
3.3	1.01292	1.87	7.3	1.02901	4.14
3.4	1.01332	1.93	7.4	1.02942	4.20
3.5	1.01371	1.99	7.5	1.02983	4.25
3.6	1.01411	2.04	7.6	1.03024	4.31
3.7	1.01451	2.10	7.7	1.03064	4.37
3.8	1.01491	2.16	7.8	1.03105	4.42
3.9	1.01531	2.21	7.9	1.03146	4.48

Degrees Brix.	Specific Gravity.	Degrees Baumé.	Degrees Brix.	Specific Gravity.	Degrees Baumé.
8.0	1.03187	4.53	13.0	1.05276	7.36
8.1	1.03228	4.59	13.1	1.05318	7.41
8.2	1.03270	4.65	13.2	1.05361	7.47
8.3	1.03311	4.70	13.3	1.05404	7.53
8.4	1.03352	4.76	13.4	1.05446	7.58
8.5	1.03393	4.82	13.5	1.05489	7.64
8.6	1.03434	4.87	13.6	1.05532	7.69
8.7	1.03475	4.93	13.7	1.05574	7.75
8.8	1.03517	4.99	13.8	1.05617	7.81
8.9	1.03558	5.04	13.9	1.05660	7.86
9.0	1.03599	5.10	14.0	1.05703	7.92
9.1	1.03640	5.16	14.1	1.05746	7.98
9.2	1.03682	5.21	14.2	1.05789	8.03
9.3	1.03723	5.27	14.3	1.05831	8.09
9.4	1.03765	5.33	14.4	1.05874	8.14
9.5	1.03806	5.38	14.5	1.05917	8.20
9.6	1.03848	5.44	14.6	1.05960	8.26
9.7	1.03889	5.50	14.7	1.06003	8.31
9.8	1.03931	5.55	14.8	1.06047	8.37
9.9	1.03972	5.61	14.9	1.06090	8.43
10.0	1.04014	5.67	15.0	1.06133	8.48
10.1	1.04055	5.72	15.1	1.06176	8.54
10.2	1.04097	5.78	15.2	1.06219	8.59
10.3	1.04139	5.83	15.3	1.06262	8.65
10.4	1.04180	5.89	15.4	1.06306	8.71
10.5	1.04222	5.95	15.5	1.06349	8.76
10.6	1.04264	6.00	15.6	1.06392	8.82
10.7	1.04306	6.06	15.7	1.06436	8.88
10.8	1.04348	6.12	15.8	1.06479	8.93
10.9	1.04390	6.17	15.9	1.06522	8.99
11.0	1.04431	6.23	16.0	1.06566	9.04
11.1	1.04473	6.29	16.1	1.06609	9.10
11.2	1.04515	6.34	16.2	1.06653	9.16
11.3	1.04557	6.40	16.3	1.06696	9.21
11.4	1.04599	6.46	16.4	1.06740	9.27
11.5	1.04641	6.51	16.5	1.06783	9.33
11.6	1.04683	6.57	16.6	1.06827	9.38
11.7	1.04726	6.62	16.7	1.06871	9.44
11.8	1.04768	6.68	16.8	1.06914	9.49
11.9	1.04810	6.74	16.9	1.06958	9.55
12.0	1.04852	6.79	17.0	1.07002	9.61
12.1	1.04894	6.85	17.1	1.07046	9.66
12.2	1.04937	6.91	17.2	1.07090	9.72
12.3	1.04979	6.96	17.3	1.07133	9.77
12.4	1.05021	7.02	17.4	1.07177	9.83
12.5	1.05064	7.08	17.5	1.07221	9.89
12.6	1.05106	7.13	17.6	1.07265	9.94
12.7	1.05149	7.19	17.7	1.07309	10.00
12.8	1.05191	7.24	17.8	1.07358	10.06
12.9	1.05233	7.30	17.9	1.07397	10.11

Degrees Brix.	Specific Gravity.	Degrees Baumé.	Degrees Brix.	Specific Gravity.	Degrees Baumé.
18.0	1.07441	10.17	23.0	1.09686	12.96
18.1	1.07485	10.22	23.1	1.09732	13.02
18.2	1.07530	10.28	23.2	1.09777	13.07
18.3	1.07574	10.33	23.3	1.09823	13.13
18.4	1.07618	10.39	23.4	1.09869	13.19
18.5	1.07662	10.45	23.5	1.09915	13.24
18.6	1.07706	10.50	23.6	1.09961	13.30
18.7	1.07751	10.56	23.7	1.10007	13.35
18.8	1.07795	10.62	23.8	1.10053	13.41
18.9	1.07839	10.67	23.9	1.10099	13.46
19.0	1.07884	10.73	24.0	1.10145	13.52
19.1	1.07928	10.78	24.1	1.10191	13.58
19.2	1.07973	10.84	24.2	1.10237	13.63
19.3	1.08017	10.90	24.3	1.10283	13.69
19.4	1.08062	10.95	24.4	1.10329	13.74
19.5	1.08106	11.01	24.5	1.10375	13.80
19.6	1.08151	11.06	24.6	1.10421	13.85
19.7	1.08196	11.12	24.7	1.10468	13.91
19.8	1.08240	11.18	24.8	1.10514	13.96
19.9	1.08285	11.23	24.9	1.10560	14.02
20.0	1.08329	11.29	25.0	1.10607	14.08
20.1	1.08374	11.34	25.1	1.10653	14.13
20.2	1.08419	11.40	25.2	1.10700	14.19
20.3	1.08464	11.45	25.3	1.10746	14.24
20.4	1.08509	11.51	25.4	1.10793	14.30
20.5	1.08553	11.57	25.5	1.10839	14.35
20.6	1.08599	11.62	25.6	1.10886	14.41
20.7	1.08643	11.68	25.7	1.10932	14.47
20.8	1.08688	11.73·	25.8	1.10979	14.52
20.9	.1.08733	11.79	25.9	1.11026	14.58
21.0	1.08778	11.85	26.0	1.11072	14.63
21.1	1.08824	11.90	26.1	1.11119	14.69
21.2	1.08869	11.96	26.2	1.11166	14.74
21.3	1.08914	12.01	26.3	1.11213	14.80
21.4	1.08959	12.07	26.4	1.11259	14.85
21.5	1.09004	12.13	26.5	1.11306	14.91
21.6	1.09049	12.18	26.6	1.11353	14.97
21.7	1.09095	12.24	26.7	1.11400	15.02
21.8	1.09140	12.29	26.8	1.11447	15.08
21.9	1.09185	12.35	26.9	1.11494	15.13
22.0	1.09231	12.40	27.0	1.11541	15.19
22.1	1.09276	12.46	27.1	1.11588	15.24
22.2	1.09321	12.52	27.2	1.11635	15.30
22.3	1.09367	12.57	27.3	1.11682	15.35
22.4	1.09412	12.63	27.4	1.11729	15.41
22.5	1.09458	12.68	27.5	1.11776	15.46
22.6	1.09503	12.74	27.6	1.11824	15.52
22.7	1.09549	12.80	27.7	1.11871	15.58
22.8	1.09595	12.85	27.8	1.11918	15.63
22.9	1.09640	12.91	27.9	1.11965	15.69

Degrees Brix.	Specific Gravity.	Degrees Baumé.	Degrees Brix.	Specific Gravity.	Degrees Baumé.
28.0	1.12013	15.74	33.0	1.14423	18.50
28.1	1.12060	15.80	33.1	1.14472	18.56
28.2	1.12107	15.85	33.2	1.14521	18.61
28.3	1.12155	15.91	33.3	1.14570	18.67
28.4	1.12202	15.96	33.4	1.14620	18.72
28.5	1.12250	16.02	33.5	1.14669	18.78
28.6	1.12297	16.07	33.6	1.14718	18.83
28.7	1.12345	16.13	33.7	1.14767	18.89
28.8	1.12393	16.18	33.8	1.14817	18.94
28.9	1.12440	16.24	33.9	1.14866	19.00
29.0	1.12488	16.30	34.0	1.14915	19.05
29.1	1.12536	16.35	34.1	1.14965	19.11
29.2	1.12583	16.41	34.2	1.15014	19.16
29.3	1.12631	16.46	34.3	1.15064	19.22
29.4	1.12679	16.52	34.4	1.15113	19.27
29.5	1.12727	16.57	34.5	1.15163	19.33
29.6	1.12775	16.63	34.6	1.15213	19.38
29.7	1.12823	16.68	34.7	1.15262	19.44
29.8	1.12871	16.74	34.8	1.15312	19.49
29.9	1.12919	16.79	34.9	1.15362	19.55
30.0	1.12967	16.85	35.0	1.15411	19.60
30.1	1.13015	16.90	35.1	1.15461	19.66
30.2	1.13063	16.96	35.2	1.15511	19.71
30.3	1.13111	17.01	35.3	1.15561	19.76
30.4	1.13159	17.07	35.4	1.15611	19.82
30.5	1.13207	17.12	35.5	1.15661	19.87
30.6	1.13255	17.18	35.6	1.15710	19.93
30.7	1.13304	17.23	35.7	1.15760	19.98
30.8	1.13352	17.29	35.8	1.15810	20.04
30.9	1.13400	17.35	35.9	1.15861	20.09
31.0	1.13449	17.40	36.0	1.15911	20.15
31.1	1.13497	17.46	36.1	1.15961	20.20
31.2	1.13545	17.51	36.2	1.16011	20.26
31.3	1.13594	17.57	36.3	1.16061	20.31
31.4	1.13642	17.62	36.4	1.16111	20.37
31.5	1.13691	17.68	36.5	1.16162	20.42
31.6	1.13740	17.73	36.6	1.16212	20.48
31.7	1.13788	17.79	36.7	1.16262	20.53
31.8	1.13837	17.84	36.8	1.16313	20.59
31.9	1.13885	17.90	36.9	1.16363	20.64
32.0	1.13934	17.95	37.0	1.16413	20.70
32.1	1.13983	18.01	37.1	1.16464	20.75
32.2	1.14032	18.06	37.2	1.16514	20.80
32.3	1.14081	18.12	37.3	1.16565	20.86
32.4	1.14129	18.17	37.4	1.16616	20.91
32.5	1.14178	18.23	37.5	1.16666	20.97
32.6	1.14227	18.28	37.6	1.16717	21.02
32.7	1.14276	18.34	37.7	1.16768	21.08
32.8	1.14325	18.39	37.8	1.16818	21.13
32.9	1.14374	18.45	37.9	1.16869	21.19

Degrees Brix.	Specific Gravity.	Degrees Baumé.	Degrees Brix.	Specific Gravity.	Degrees Baumé.
38.0	1.16920	21.24	43.0	1.19505	23.96
38.1	1.16971	21.30	43.1	1.19558	24.01
38.2	1.17022	21.35	43.2	1.19611	24.07
38.3	1.17072	21.40	43.3	1.19663	24.12
38.4	1.17132	21.46	43.4	1.19716	24.17
38.5	1.17174	21.51	43.5	1.19769	24.23
38.6	1.17225	21.57	43.6	1.19822	24.28
38.7	1.17276	21.62	43.7	1.19875	24.34
38.8	1.17327	21.68	43.8	1.19927	24.39
38.9	1.17379	21.73	43.9	1.19980	24.44
39.0	1.17430	21.79	44.0	1.20033	24.50
39.1	1.17481	21.84	44.1	1.20086	24.55
39.2	1.17532	21.90	44.2	1.20139	24.61
39.3	1.17583	21.95	44.3	1.20192	24.66
39.4	1.17635	22.00	44.4	1.20245	24.71
39.5	1.17686	22.06	44.5	1.20299	24.77
39.6	1.17737	22.11	44.6	1.20352	24.82
39.7	1.17789	22.17	44.7	1.20405	24.88
39.8	1.17840	22.22	44.8	1.20458	24.93
39.9	1.17892	22.28	44.9	1.20512	24.98
40.0	1.17943	22.33	45.0	1.20565	25.04
40.1	1.17995	22.38	45.1	1.20618	25.09
40.2	1.18046	22.44	45.2	1.20672	25.14
40.3	1.18098	22.49	45.3	1.20725	25.20
40.4	1.18150	22.55	45.4	1.20779	25.25
40.5	1.18201	22.60	45 5	1.20832	25.31
40.6	1.18253	22.66	45.6	1.20886	25.36
40.7	1.18305	22.71	45.7	1.20939	25.41
40.8	1.18357	22.77	45.8	1.20993	25.47
40.9	1.18408	22.82	45.9	1.21046	25.52
41.0	1.18460	22.87	46.0	1.21100	25.57
41.1	1.18512	22.93	46.1	1.21154	25.63
41.2	1.18564	22.98	46.2	1.21208	25.68
41.3	1.18616	23.04	46.3	1.21261	25.74
41.4	1.18668	23.09	46.4	1.21315	25.79
41.5	1.18720	23.15	46.5	1.21369	25.84
41.6	1.18772	23.20	46.6	1.21423	25.90
41.7	1.18824	23.25	46.7	1.21477	25.95
41.8	1.18877	23.31	46.8	1.21531	26.00
41.9	1.18929	23.36	46.9	1.21585	26.06
42.0	1.18981	23.42	47.0	1.21639	26.11
42.1	1.19033	23.47	47.1	1.21693	26.17
42.2	1.19086	23.52	47.2	1.21747	26.22
42.3	1.19138	23.58	47.3	1.21802	26.27
42.4	1.19190	23.63	47.4	1.21856	26.33
42.5	1.19243	23.69	47.5	1.21910	26.38
42.6	1.19295	23.74	47.6	1.21964	26.43
42.7	1.19348	23.79	47.7	1.22019	26.49
42.8	1.19400	23.85	47.8	1.22073	26.54
42.9	1.19453	23.90	47.9	1.22127	26.59

Degrees Brix.	Specific Gravity.	Degrees Baumé.	Degrees Brix.	Specific Gravity.	Degrees Baumé.
48.0	1.22182	26.65	53.0	1.24951	29.31
48.1	1.22236	26.70	53.1	1.25008	29.36
48.2	1.22291	26.75	53.2	1.25064	29.42
48.3	1.22345	26.81	53.3	1.25120	29.47
48.4	1.22400	26.86	53.4	1.25177	29.52
48.5	1.22455	26.92	53.5	1.25233	29.57
48.6	1.22509	26.97	53.6	1.25290	29.63
48.7	1.22564	27.02	53.7	1.25347	29.68
48.8	1.22619	27.08	53.8	1.25403	29.73
48.9	1.22673	27.13	53.9	1.25460	29.79
49.0	1.22728	27.18	54.0	1.25517	29.84
49.1	1.22783	27.24	54.1	1.25573	29.89
49.2	1.22838	27.29	54.2	1.25630	29.94
49.3	1.22893	27.34	54.3	1.25687	30.00
49.4	1.22948	27.40	54.4	1.25744	30.05
49.5	1.23003	27.45	54.5	1.25801	30.10
49.6	1.23058	27.50	54.6	1.25857	30.16
49.7	1.23113	27.56	54.7	1.25914	30.21
49.8	1.23168	27.61	54.8	1.25971	30.26
49.9	1.23223	27.66	54.9	1.26028	30.31
50.0	1.23278	27.72	55.0	1.26086	30.37
50.1	1.23334	27.77	55.1	1.26143	30.42
50.2	1.23389	27.82	55.2	1.26200	30.47
50.3	1.23444	27.88	55.3	1.26257	30.53
50.4	1.23499	27.93	55.4	1.26314	30.58
50.5	1.23555	27.98	55.5	1.26372	30.63
50.6	1.23610	28.04	55.6	1.26429	30.68
50.7	1.23666	28.09	55.7	1.26486	30.74
50.8	1.23721	28.14	55.8	1.26544	30.79
50.9	1.23777	28.20	55.9	1.26601	30.84
51.0	1.23832	28.25	56.0	1.26658	30.89
51.1	1.23888	28.30	56.1	1.26716	30.95
51.2	1.23943	28.36	56.2	1.26773	31.00
51.3	1.23999	28.41	56.3	1.26831	31.05
51.4	1.24055	28.46	56.4	1.26889	31.10
51.5	1.24111	28.51	56.5	1.26946	31.16
51.6	1.24166	28.57	56.6	1.27004	31.21
51.7	1.24222	28.62	56.7	1.27062	31.26
51.8	1.24278	28.67	56.8	1.27120	31.31
51.9	1.24334	28.73	56.9	1.27177	31.37
52.0	1.24390	28.78	57.0	1.27235	31.42
52.1	1.24446	28.83	57.1	1.27293	31.47
52.2	1.24502	28.89	57.2	1.27351	31.52
52.3	1.24558	28.94	57.3	1.27409	31.58
52.4	1.24614	28.99	57.4	1.27467	31.63
52.5	1.24670	29.05	57.5	1.27525	31.68
52.6	1.24726	29.10	57.6	1.27583	31.73
52.7	1.24782	29.15	57.7	1.27641	31.79
52.8	1.24839	29.20	57.8	1.27699	31.84
52.9	1.24895	29.26	57.9	1.27758	31.89

Degrees Brix.	Specific Gravity.	Degrees Baumé.	Degrees Brix.	Specific Gravity.	Degrees Baumé.
58.0	1.27816	31.94	63.0	1.30777	34.54
58.1	1.27874	32.00	63.1	1.30837	34.59
58.2	1.27932	32.05	63.2	1.30897	34.65
58.3	1.27991	32.10	63.3	1.30958	34.70
58.4	1.28049	32.15	63.4	1.31018	34.75
58.5	1.28107	32.20	63.5	1.31078	34.80
58.6	1.28166	32.26	63.6	1.31139	34.85
58.7	1.28224	32.31	63.7	1.31199	34.90
58.8	1.28283	32.36	63.8	1.31260	34.96
58.9	1.28342	32.41	63.9	1.31320	35.01
59.0	1.28400	32.47	64.0	1.31381	35.06
59.1	1.28459	32.52	64.1	1.31442	35.11
59.2	1.28518	32.57	64.2	1.31502	35.16
59.3	1.28576	32.62	64.3	1.31563	35.21
59.4	1.28635	32.67	64.4	1.31624	35.27
59.5	1.28694	32.73	64.5	1.31684	35.32
59.6	1.28753	32.78	64.6	1.31745	35.37
59.7	1.28812	32.83	64.7	1.31806	35.42
59.8	1.28871	32.88	64.8	1.31867	35.47
59.9	1.28930	32.93	64.9	1.31928	35.52
60.0	1.28989	32.99	65.0	1.31989	35.57
60.1	1.29048	33.04	65.1	1.32050	35.63
60.2	1.29107	33.09	65.2	1.32111	35.68
60.3	1.29166	33.14	65.3	1.32172	35.73
60.4	1.29225	33.20	65.4	1.32233	35.78
60.5	1.29284	33.25	65.5	1.32294	35.83
60.6	1.29343	33.30	65.6	1.32355	35.88
60.7	1.29403	33.35	65.7	1.32417	35.93
60.8	1.29462	33.40	65.8	1.32478	35.98
60.9	1.29521	33.46	65.9	1.32539	36.04
61.0	1.29581	33.51	66.0	1.32601	36.09
61.1	1.29640	33.56	66.1	1.32662	36.14
61.2	1.29700	33.61	66.2	1.32724	36.19
61.3	1.29759	33.66	66.3	1.32785	36.24
61.4	1.29819	33.71	66.4	1.32847	36.29
61.5	1.29878	33.77	66.5	1.32908	36.34
61.6	1.29938	33.82	66.6	1.32970	36.39
61.7	1.29998	33.87	66.7	1.33031	36.45
61.8	1.30057	33.92	66.8	1.33093	36.50
61.9	1.30117	33.97	66.9	1.33155	36.55
62.0	1.30177	34.03	67.0	1.33217	36.60
62.1	1.30237	34.08	67.1	1.33278	36.65
62.2	1.30297	34.13	67.2	1.33340	36.70
62.3	1.30356	34.18	67.3	1.33402	36.75
62.4	1.30416	34.23	67.4	1.33464	36.80
62.5	1.30476	34.28	67.5	1.33526	36.85
62.6	1.30536	34.34	67.6	1.33588	36.90
62.7	1.30596	34.39	67.7	1.33650	36.96
62.8	1.30657	34.44	67.8	1.33712	37.01
62.9	1.30717	34.49	67.9	1.33774	37.06

Degrees Brix.	Specific Gravity.	Degrees Baumé.	Degrees Brix.	Specific Gravity.	Degrees Baumé.
68.0	1.33836	37.11	73.0	1.36995	39.64
68.1	1.33899	37.16	73.1	1.37059	39.69
68.2	1.33961	37.21	73.2	1.37124	39.74
68.3	1.34023	37.26	73.3	1.37188	39.79
68.4	1.34085	37.31	73.4	1.37252	39.84
68.5	1.34148	37.36	73.5	1.37317	39.89
68.6	1.34210	37.41	73.6	1.37381	39.94
68.7	1.34273	37.47	73.7	1.37446	39.99
68.8	1.34335	37.52	73.8	1.37510	40.04
68.9	1.34398	37.57	73.9	1.37575	40.09
69.0	1.34460	37.62	74.0	1.37639	40.14
69.1	1.34523	37.67	74.1	1.37704	40.19
69.2	1.34585	37.72	74.2	1.37768	40.24
69.3	1.34648	37.77	74.3	1.37833	40.29
69.4	1.34711	37.82	74.4	1.37898	40.34
69.5	1.34774	37.87	74.5	1.37962	40.39
69.6	1.34836	37.92	74.6	1.38027	40.44
69.7	1.34899	37.97	74.7	1.38092	40.49
69.8	1.34962	38.02	74.8	1.38157	40.54
69.9	1.35025	38.07	74.9	1.38222	40.59
70.0	1.35088	38.12	75.0	1.38287	40.64
70.1	1.35151	38.18	75.1	1.38352	40.69
70.2	1.35214	38.23	75.2	1.38417	40.74
70.3	1.35277	38.28	75.3	1.38482	40.79
70.4	1.35340	38.33	75.4	1.38547	40.84
70.5	1.35403	38.38	75.5	1.38612	40.89
70.6	1.35466	38.43	75.6	1.38677	40 94
70.7	1.35530	38.48	75.7	1.38743	40.99
70.8	1.35593	38.53	75.8	1.38808	41.04
70.9	1.35656	38.58	75.9	1.38873	41.09
71.0	1.35720	38.63	76.0	1.38939	41.14
71.1	1.35783	38.68	76.1	1.39004	41.19
71.2	1.35847	38.73	76.2	1.39070	41.24
71.3	1.35910	38.78	76.3	1.39135	41.29
71.4	1.35974	38.83	76.4	1.39201	41.33
71.5	1.36037	38.88	76.5	1.39266	41.38
71.6	1.36101	38.93	76.6	1.39332	41.43
71.7	1.36164	38.98	76.7	1.39397	41.48
71.8	1.36228	39.03	76.8	1.39463	41.53
71.9	1.36292	39.08	76.9	1.39529	41.58
72.0	1.36355	39.13	77.0	1.39595	41.63
72.1	1.36419	39.19	77.1	1.39660	41.68
72.2	1.36483	39.24	77.2	1.39726	41.73
72.3	1.36547	39.29	77.3	1.39792	41.78
72.4	1.36611	39.34	77.4	1.39858	41.83
72.5	1.36675	39 39	77.5	1.39924	41.88
72.6	1.36739	39.44	77.6	1.39990	41.93
72.7	1.36803	39.49	77.7	1.40056	41.98
72.8	1.36867	39.54	77.8	1.40122	42.03
72.9	1.36931	39.59	77.9	1.40188	42.08

Degrees Brix.	Specific Gravity.	Degrees Baumé.	Degrees Brix.	Specific Gravity.	Degrees Baumé.
78.0	1.40254	42.13	83.0	1.43614	44.58
78.1	1.40321	42.18	83.1	1.43682	44.62
78.2	1.40387	42.23	83.2	1.43750	44.67
78.3	1.40453	42.28	83.3	1.43819	44.72
78.4	1.40520	42.32	83.4	1.43887	44.77
78.5	1.40586	42.37	83.5	1.43955	44.82
78.6	1.40652	42.42	83.6	1.44024	44.87
78.7	1.40719	42.47	83.7	1.44092	44.91
78.8	1.40785	42.52	83.8	1.44161	44.96
78.9	1.40852	42.57	83.9	1.44229	45.01
79.0	1.40918	42.62	84.0	1.44298	45.06
79.1	1.40985	42.67	84.1	1.44367	45.11
79.2	1.41052	42.72	84.2	1.44435	45.16
79.3	1.41118	42.77	84.3	1.44504	45.21
79.4	1.41185	42.82	84.4	1.44573	45.25
79.5	1.41252	42.87	84.5	1.44641	45.30
79.6	1.41318	42.92	84.6	1.44710	45.35
79.7	1.41385	42.96	84.7	1.44779	45.40
79.8	1.41452	43.01	84.8	1.44848	45.45
79.9	1.41519	43.06	84.9	1.44917	45.49
80.0	1.41586	43.11	85.0	1.44986	45.54
80.1	1.41653	43.16	85.1	1.45055	45.59
80.2	1.41720	43.21	85.2	1.45124	45.64
80.3	1.41787	43.26	85.3	1.45193	45.69
80.4	1.41854	43.31	85.4	1.45262	45.74
80.5	1.41921	43.36	85.5	1.45331	45.78
80.6	1.41989	43.41	85.6	1.45401	45.83
80.7	1.42056	43.45	85.7	1.45470	45.88
80.8	1.42123	43.50	85.8	1.45539	45.93
80.9	1.42190	43.55	85.9	1.45609	45.98
81.0	1.42253	43.60	86.0	1.45678	46.02
81.1	1.42325	43.65	86.1	1.45748	46.07
81.2	1.42393	43.70	86.2	1.45817	46.12
81.3	1.42460	43.75	86.3	1.45887	46.17
81.4	1.42528	43.80	86.4	1.45956	46.22
81.5	1.42595	43.85	86.5	1.46026	46.26
81.6	1.42663	43.89	86.6	1.46095	46.31
81.7	1.42731	43.94	86.7	1.46165	46.36
81.8	1.42798	43.99	86.8	1.46235	46.41
81.9	1.42866	44.04	86.9	1.46304	46.46
82.0	1.42934	44.09	87.0	1.46374	46.50
82.1	1.43002	44.14	87.1	1.46444	46.55
82.2	1.43070	44.19	87.2	1.46514	46.60
82.3	1.43137	44.24	87.3	1.46584	46.65
82.4	1.43205	44.28	87.4	1.46654	46.69
82.5	1.43273	44.33	87.5	1.46724	46.74
82.6	1.43341	44.38	87.6	1.46794	46.79
82.7	1.43409	44.43	87.7	1.46864	46.84
82.8	1.43478	44.48	87.8	1.46934	46.88
82.9	1.43546	44.53	87.9	1.47004	46.93

Degrees Brix.	Specific Gravity.	Degrees Baumé.	Degrees Brix.	Specific Gravity.	Degrees Baumé.
88.0	1.47074	46.98	93.0	1.50635	49.34
88.1	1.47145	47.03	93.1	1.50707	49.39
88.2	1.47215	47.08	93.2	1.50779	49.43
88.3	1.47285	47.12	93.3	1.50852	49.48
88.4	1.47356	47.17	93.4	1.50924	49.53
88.5	1.47426	47.22	93.5	1.50996	49.57
88.6	1.47496	47.27	93.6	1.51069	49.62
88.7	1.47567	47.31	93.7	1.51141	49.67
88.8	1.47637	47.36	93.8	1.51214	49.71
88.9	1.47708	47.41	93.9	1.51286	49.76
89.0	1.47778	47.46	94.0	1.51359	49.81
89.1	1.47849	47.50	94.1	1.51431	49.85
89.2	1.47920	47.55	94.2	1.51504	49.90
89.3	1.47991	47.60	94.3	1.51577	49.94
89.4	1.48061	47.65	94.4	1.51649	49.99
89.5	1.48132	47.69	94.5	1.51722	50.04
89.6	1.48203	47.74	94.6	1.51795	50.08
89.7	1.48274	47.79	94.7	1.51868	50.13
89.8	1.48345	47.83	94.8	1.51941	50.18
89.9	1.48416	47.88	94.9	1.52014	50.22
90.0	1.48486	47.93	95.0	1.52087	50.27
90.1	1.48558	47.98	95.1	1.52159	50.32
90.2	1.48629	48.02	95.2	1.52232	50.36
90.3	1.48700	48.07	95.3	1.52304	50.41
90.4	1.48771	48.12	95.4	1.52376	50.45
90.5	1.48842	48.17	95.5	1.52449	50.50
90.6	1.48913	48.21	95.6	1.52521	50.55
90.7	1.48985	48.26	95.7	1.52593	50.59
90.8	1.49056	48.31	95.8	1.52665	50.64
90.9	1.49127	48.35	95.9	1.52738	50.69
91.0	1.49199	48.40	96.0	1.52810	50.73
91.1	1.49270	48.45	96.1	1.52884	50.78
91.2	1.49342	48.50	96.2	1.52958	50.82
91.3	1.49413	48.54	96.3	1.53032	50.87
91.4	1.49485	48.59	96.4	1.53106	50.92
91.5	1.49556	48.64	96.5	1.53180	50.96
91.6	1.49628	48.68	96.6	1.53254	51.01
91.7	1.49700	48.73	96.7	1.53328	51.05
91.8	1.49771	48.78	96.8	1.53402	51.10
91.9	1.49843	48.82	96.9	1.53476	51.15
92.0	1.49915	48.87	97.0	1.53550	51.19
92.1	1.49987	48.92	97.1	1.53624	51.24
92.2	1.50058	48.96	97.2	1.53698	51.28
92.3	1.50130	49.01	97.3	1.53772	51.33
92.4	1.50202	49.06	97.4	1.53846	51.38
92.5	1.50274	49.11	97.5	1.53920	51.42
92.6	1.50346	49.15	97.6	1.53994	51.47
92.7	1.50419	49.20	97.7	1.54068	51.51
92.8	1.50491	49.25	97.8	1.54142	51.56
92.9	1.50563	49.29	97.9	1.54216	51.60

Degrees Brix.	Specific Gravity.	Degrees Baumé.	Degrees Brix.	Specific Gravity.	Degrees Baumé.
98.0	1.54290	51.65	99.0	1.55040	52.11
98.1	1.54365	51.70	99.1	1.55115	52.15
98.2	1.54440	51.74	99.2	1.55189	52.20
98.3	1.54515	51.79	99.3	1.55264	52.24
98.4	1.54590	51.83	99.4	1.55338	52.29
98.5	1.54665	51.88	99.5	1.55413	52.33
98.6	1.54740	51.92	99.6	1.55487	52.38
98.7	1.54815	51.97	99.7	1.55562	52.42
98.8	1.54890	52.01	99.8	1.55636	52.47
98.9	1.54965	52.06	99.9	1.55711	52.51
			100.0	1.55785	52.56

II.

CORRECTIONS FOR TEMPERATURE IN DE-TERMINATIONS BY THE SPECIFIC GRAV-ITY HYDROMETER.

(CASAMAJOR.)

II.

Normal Temperature : 15.0° C.		Normal Temperature : 17.5° C.	
Temperature in Degrees Centigrade.	Add to the Reading of the Hydrometer.	Temperature in Degrees Centigrade.	Add to the Reading of the Hydrometer.
9.90	—0.0005	7.5	—0.0010
15.00	0.0000	13.0	—0.0005
18.20	+0.0005	17.5	0.0000
20.75	0.0010	20.2	+0.0005
23.20	0.0015	23.0	0.0010
25.30	0.0020	25.0	0.0015
27.30	0.0025	27.0	0.0020
29.40	0.0030	29.0	0.0025
31.20	0.0035	31.0	0.0030
32.80	0.0040	32.5	0.0035
34.50	0.0045	34.7	0.0040
36.10	0.0050	36.2	0.0045
37.60	0.0055	37.4	0.0050
38.80	0.0060	39.0	0.0055
40.40	0.0065	40.5	0.0060
41.60	0.0070	42.0	0.0065
42.90	0.0075	43.4	0.0070
44.20	0.0080	44.2	0.0075
45.00	0.0083	45.0	0.0080

III.

CORRECTIONS FOR TEMPERATURE IN DETERMINATIONS BY THE BRIX HYDROMETER.

Normal Temperature = 17.5° C.

(STAMMER.)

181

III.

Degree Centigrade.	Degree Brix of the Solution.												
	0	5	10	15	20	25	30	35	40	50	60	70	75
	The degree read is to be *decreased* by—												
0°	0.17	0.30	0.41	0.52	0.62	0.72	0.82	0.92	0.98	1.11	1.22	1.25	1.29
5	0.23	0.30	0.37	0.44	0.52	0.59	0.65	0.72	0.75	0.80	0.88	0.91	0.94
10	0.20	0.26	0.29	0.33	0.36	0.39	0.42	0.45	0.48	0.50	0.54	0.58	0.61
11	0.18	0.23	0.26	0.28	0.31	0.34	0.36	0.39	0.41	0.43	0.47	0.50	0.53
12	0.16	0.20	0.22	0.24	0.26	0.29	0.31	0.33	0.34	0.36	0.40	0.42	0.46
13	0.14	0.18	0.19	0.21	0.22	0.24	0.26	0.27	0.28	0.29	0.33	0.35	0.39
14	0.12	0.15	0.16	0.17	0.18	0.19	0.21	0.22	0.22	0.23	0.26	0.28	0.32
15	0.09	0.11	0.12	0.14	0.14	0.15	0.16	0.17	0.16	0.17	0.19	0.21	0.25
16	0.06	0.07	0.08	0.09	0.10	0.10	0.11	0.12	0.12	0.12	0.14	0.16	0.18
17	0.02	0.02	0.03	0.03	0.03	0.04	0.04	0.04	0.04	0.04	0.05	0.05	0.06
	The degree read is to be *increased* by—												
18	0.02	0.03	0.03	0.03	0.03	0.03	0.03	0.03	0.03	0.03	0.03	0.03	0.02
19	0.06	0.08	0.08	0.09	0.09	0.10	0.10	0.10	0.10	0.10	0.10	0.08	0.06
20	0.11	0.14	0.15	0.17	0.17	0.18	0.18	0.18	0.19	0.19	0.18	0.15	0.11
21	0.16	0.20	0.22	0.24	0.24	0.25	0.25	0.25	0.26	0.26	0.25	0.22	0.18
22	0.21	0.26	0.29	0.31	0.31	0.32	0.32	0.32	0.33	0.34	0.32	0.29	0.25
23	0.27	0.32	0.35	0.37	0.38	0.39	0.39	0.39	0.40	0.42	0.39	0.36	0.33
24	0.32	0.38	0.41	0.43	0.44	0.46	0.46	0.47	0.47	0.50	0.46	0.43	0.40
25	0.37	0.44	0.47	0.49	0.51	0.53	0.54	0.55	0.55	0.58	0.54	0.51	0.48
26	0.43	0.50	0.54	0.56	0.58	0.60	0.61	0.62	0.62	0.66	0.62	0.58	0.55
27	0.49	0.57	0.61	0.63	0.65	0.68	0.68	0.69	0.70	0.74	0.70	0.65	0.62
28	0.56	0.64	0.68	0.70	0.72	0.76	0.76	0.78	0.78	0.82	0.78	0.72	0.70
29	0.63	0.71	0.75	0.78	0.79	0.84	0.84	0.86	0.86	0.90	0.86	0.80	0.78
30	0.70	0.78	0.82	0.87	0.87	0.92	0.92	0.94	0.94	0.98	0.94	0.88	0.86
35	1.10	1.17	1.22	1.24	1.30	1.32	1.33	1.35	1.36	1.39	1.34	1.27	1.25
40	1.50	1.61	1.67	1.71	1.73	1.79	1.79	1.80	1.82	1.83	1.78	1.69	1.65
50	....	2.65	2.71	2.74	2.78	2.80	2.80	2.80	2.80	2.79	2.70	2.56	2.51
60	...	3.87	3.88	3.88	3.88	3.88	3.88	3.88	3.90	3.82	3.70	3.43	3.41
70	....	....	5.18	5.20	5.14	5.13	5.10	5.08	5.06	4.90	4.72	4.47	4.35
80	....	....	6.62	6.59	6.54	6.46	6.38	6.30	6.26	6.06	5.82	5.50	5.33

IV.

FACTORS.

Arranged for Specific Gravity Determinations.

Calculated for Wiechmann: Sugar Analysis, from the data given in Table I.

$$\text{Factor} = \frac{26.048}{\text{Degree Brix} \times \text{Specific Gravity}}.$$

IV.

Specific Gravity.	Factor.	Specific Gravity.	Factor.	Specific Gravity.	Factor.	Specific Gravity.	Factor.
1.0950	1.053	1.0980	1.023	1.1010	0.990	1.1040	0.959
1.0955	1.047	1.0985	1.013	1.1015	0.985	1.1045	0.955
1.0960	1.042	1.0990	1.008	1.1020	0.981	1.1050	0.950
1.0965	1.037	1.0995	1.004	1.1025	0.976	1.1055	0.946
1.0970	1.033	1.1000	1.000	1.1030	0.972	1.1060	0.942
1.0975	1.028	1.1005	0.944	1.1035	0.968		

V.

FACTORS.

Arranged for Brix determinations.

Calculated for Wiechmann: Sugar Analysis, from the data given in Table I.

$$\text{Factor} = \frac{26.048}{\text{Degree Brix} \times \text{Specific Gravity}}.$$

V.

Degree Brix.	0	1	2	3	4	5	6	7	8	9
0	.....	260.381	130.140	86.726	65.019	51.996	43.313	37.111	32.459	28.842
1	25.947	23.579	21.606	19.936	18.505	17.265	16.179	15.222	14.370	13.609
2	12.923	12.303	11.739	11.225	10.753	10.318	9.918	9.547	9.202	8.881
3	8.582	8.302	8.039	7.793	7.560	7.342	7.135	6.939	6.754	6.578
4	6.411	6.253	6.101	5.957	5.819	5.688	5.562	5.441	5.326	5.215
5	5.109	5.007	4.909	4.814	4.723	4.635	4.551	4.469	4.390	4.314
6	4.241	4.170	4.101	4.034	3.969	3.907	3.846	3.787	3.730	3.674
7	3.621	3.568	3.517	3.468	3.419	3.372	3.327	3.282	3.239	3.197
8	3.155	3.115	3.076	3.038	3.000	2.964	2.928	2.893	2.859	2.826
9	2.794	2.762	2.731	2.700	2.671	2.641	2.613	2.585	2.557	2.531
10	2.504	2.479	2.453	2.428	2.404	2.380	2.357	2.334	2.311	2.289
11	2.268	2.246	2.225	2.205	2.185	2.165	2.145	2.126	2.107	2.088
12	2.070	2.052	2.035	2.017	2.000	1.983	1.967	1.951	1.935	1.919
13	1.903	1.888	1.873	1.858	1.843	1.829	1.815	1.801	1.787	1.774
14	1.760	1.747	1.734	1.721	1.709	1.696	1.684	1.672	1.660	1.648
15	1.636	1.625	1.613	1.602	1.591	1.580	1.569	1.559	1.548	1.538
16	1.528	1.518	1.508	1.498	1.488	1.478	1.469	1.459	1.450	1.441
17	1.432	1.423	1.414	1.405	1.397	1.388	1.380	1.371	1.363	1.355
18	1.347	1.339	1.331	1.323	1.315	1.308	1.300	1.293	1.285	1.278
19	1.271	1.264	1.256	1.249	1.243	1.236	1.229	1.222	1.215	1.209
20	1.202	1.196	1.189	1.183	1.177	1.171	1.164	1.158	1.152	1.146
21	1.140	1.134	1.129	1.123	1.117	1.111	1.106	1.100	1.095	1.089
22	1.084	1.079	1.073	1.068	1.063	1.058	1.053	1.047	1.042	1.037
23	1.033	1.028	1.023	1.018	1.013	1.008	1.004	0.999	0.994	0.990
24	0.985	0.981	0.976	0.972	0.968	0.963	0.959	0.955	0.950	0.946
25	0.942	0.938	0.934	0.930	0.926	0.922	0.918	0.914	0.910	0.906
26	0.902	0.898	0.894	0.891	0.887	0.883	0.879	0.876	0.872	0.869
27	0.865	0.861	0.858	0.854	0.851	0.847	0.844	0.841	0.837	0.834
28	0.831									

VI.

ESTIMATION OF PERCENTAGE OF SUGAR BY WEIGHT, IN WEAK SUGAR SOLUTIONS.

Tucker: Manual of Sugar Analysis.

Abridged from a table calculated by:

(OSWALD.)

VI.

Degree Brix.	Specific Gravity.	READING OF THE SACCHARIMETER.									
		1	2	3	4	5	6	7	8	9	10
0.0	1.0000	.260	.521	.781	1.042	1.302	1.563	1.823	2.084	2.344	2.605
0.5	1.0019	.260	.520	.780	1.040	1.300	1.560	1.820	2.080	2.340	2.600
1.0	1.0039	.259	.519	.778	1.038	1.297	1.557	1.816	2.076	2.335	2.595
1.5	1.0058	.259	.518	.777	1.036	1.295	1.554	1.813	2.072	2.331	2.590
2.0	1.0078	.258	.517	.775	1.034	1.292	1.551	1.809	2.068	2.326	2.585
2.5	1.0097	.258	.516	.774	1.032	1.290	1.548	1.806	2.064	2.322	2.580
3.0	1.0117	.257	.515	.772	1.029	1.287	1.545	1.802	2.060	2.317	2.575
3.5	1.0137	.257	.514	.771	1.028	1.285	1.542	1.799	2.056	2.313	2.570
4.0	1.0157	.256	.513	.769	1.026	1.282	1.539	1.795	2.052	2.308	2.565
4.5	1.0177	.256	.512	.768	1.024	1.280	1.536	1.792	2.048	2.304	2.559
5.0	1.0197	.255	.511	.766	1.022	1.277	1.533	1.788	2.044	2.299	2.554
5.5	1.0213	.255	.510	.765	1.020	1.275	1.530	1.785	2.040	2.295	2.549
6.0	1.0237	.254	.509	.763	1.018	1.272	1.527	1.781	2.036	2.290	2.544
6.5	1.0257	.254	.508	.762	1.016	1.270	1.524	1.778	2.032	2.285	2.539
7.0	1.0278	.253	.507	.760	1.014	1.267	1.521	1.774	2.027	2.281	2.534
7.5	1.0298	.253	.506	.758	1.012	1.265	1.518	1.771	2.023	2.276	2.529
8.0	1.0319	.252	.505	.757	1.010	1.262	1.515	1.767	2.019	2.272	2.524
8.5	1.0339	.252	.504	.756	1.008	1.260	1.512	1.763	2.015	2.267	2.519
9.0	1.0360	.251	.503	.754	1.006	1.257	1.509	1.760	2.011	2.263	2.514
9.5	1.0380	.251	.502	.753	1.004	1.255	1.506	1.757	2.007	2.258	2.509
10.0	1.0410	.250	.501	.751	1.002	1.252	1.503	1.753	2.003	2.254	2.504
10.5	1.0422	.250	.500	.750	1.000	1.250	1.500	1.750	1.999	2.249	2.499
11.0	1.0443	.249	.499	.748	.998	1.247	1.497	1.746	1.995	2.245	2.494
11.5	1.0464	.249	.498	.747	.996	1.245	1.494	1.743	1.991	2.240	2.489
12.0	1.0485	.248	.497	.745	.994	1.242	1.491	1.739	1.987	2.236	2.484
12.5	1.0506	.248	.496	.744	.992	1.240	1.488	1.735	1.983	2.231	2.479
13.0	1.0528	.247	.495	.742	.990	1.237	1.484	1.732	1.979	2.227	2.474
13.5	1.0549	.247	.494	.741	.988	1.235	1.482	1.728	1.975	2.222	2.469
14.0	1.0570	.246	.493	.739	.986	1.232	1.479	1.725	1.971	2.218	2.464
14.5	1.0591	.246	.492	.738	.984	1.230	1.476	1.722	1.967	2.213	2.459
15.0	1.0613	.245	.491	.736	.982	1.227	1.473	1.718	1.963	2.209	2.454
15.5	1.0635	.245	.490	.735	.980	1.225	1.470	1.714	1.959	2.204	2.449
16.0	1.0657	.244	.489	.733	.978	1.222	1.467	1.711	1.955	2.200	2.444
16.5	1.0678	.244	.488	.732	.976	1.220	1.464	1.708	1.951	2.195	2.439
17.0	1.0700	.243	.487	.730	.974	1.217	1.461	1.704	1.948	2.191	2.434
17.5	1.0722	.243	.486	.729	.972	1.215	1.458	1.701	1.944	2.186	2.429
18.0	1.0744	.242	.485	.727	.970	1.212	1.455	1.697	1.940	2.182	2.424
18.5	1.0765	.242	.484	.726	.968	1.210	1.452	1.694	1.936	2.178	2.420
19.0	1.0787	.241	.483	.724	.966	1.207	1.449	1.690	1.932	2.173	2.415
19.5	1.0810	.241	.482	.723	.964	1.205	1.446	1.687	1.928	2.169	2.410
20.0	1.0833	.240	.481	.721	.962	1.202	1.443	1.683	1.924	2.164	2.405
20.5	1.0855	.240	.480	.720	.960	1.200	1.440	1.680	1.920	2.160	2.400
21.0	1.0878	.239	.479	.718	.958	1.197	1.437	1.676	1.916	2.155	2.395
21.5	1.0900	.239	.478	.717	.956	1.195	1.434	1.673	1.912	2.151	2.390
22.0	1.0923	.238	.477	.715	.954	1.192	1.431	1.669	1.908	2.146	2.385
22.5	1.0946	.238	.476	.714	.952	1.190	1.428	1.666	1.904	2.142	2.380
23.0	1.0969	.237	.475	.712	.950	1.187	1.425	1.662	1.900	2.137	2.375

VII.

"HUNDRED POLARIZATION."

(SCHEIBLER.)

VII.

Degrees read.	Instead of 13.024 g. there must be taken.		Degrees read.	Instead of 13.024 g. there must be taken.		Degrees read.	Instead of 13.024 g. there must be taken.	
	Grammes.	Difference.		Grammes.	Difference.		Grammes.	Difference.
82.0	15.883	2.859	86.0	15.144	2.120	90.0	14.471	1.447
1	864	840	1	127	103	1	455	431
2	844	820	2	109	085	2	439	415
3	825	801	3	092	068	3	423	399
4	806	782	4	074	050	4	407	383
5	778	763	5	057	033	5	391	367
6	768	744	6	039	015	6	375	351
7	748	724	7	022	1.998	7	359	335
8	729	705	8	005	981	8	344	320
9	710	686	9	14.987	963	9	328	304
83.0	692	668	87.0	970	946	91.0	312	288
1	673	649	1	953	929	1	296	272
2	654	630	2	936	912	2	281	257
3	635	611	3	919	895	3	265	241
4	616	592	4	902	878	4	249	225
5	598	574	5	885	861	5	234	210
6	579	555	6	868	844	6	218	194
7	560	536	7	851	827	7	203	179
8	542	518	8	834	810	8	187	163
9	523	499	9	817	793	9	172	148
84.0	505	481	88.0	800	776	92.0	157	133
1	486	462	1	783	759	1	141	117
2	468	444	2	766	742	2	126	102
3	450	426	3	750	726	3	111	087
4	431	407	4	733	709	4	095	071
5	413	389	5	717	693	5	080	056
6	395	371	6	700	676	6	065	041
7	377	353	7	683	659	7	050	026
8	358	334	8	667	643	8	034	010
9	340	316	9	650	626	9	019	0.995
85.0	322	298	89.0	634	610	93.0	004	980
1	304	280	1	617	593	1	13.989	965
2	286	262	2	601	577	2	974	950
3	268	244	3	585	561	3	959	935
4	251	227	4	568	544	4	944	920
5	233	209	5	552	528	5	929	905
6	215	191	6	536	512	6	915	891
7	197	173	7	520	496	7	900	876
8	179	155	8	503	479	8	885	861
9	162	138	9	487	463	9	870	846

Degrees read.	Instead of 13.024 g. there must be taken.		Degrees read.	Instead of 13.024 g. there must be taken.		Degrees read.	Instead of 13.024 g. there must be taken.	
	Grammes.	Difference.		Grammes.	Difference.		Grammes.	Difference.
94.0	13.855	0.831	96.0	13.567	0.543	98.0	13.290	0.266
1	841	817	1	553	529	1	276	252
2	826	802	2	538	514	2	263	239
3	811	787	3	524	500	3	249	225
4	797	773	4	510	486	4	236	212
5	782	758	5	496	472	5	222	198
6	767	743	6	482	458	6	209	185
7	753	729	7	468	444	7	196	172
8	738	714	8	455	431	8	182	158
9	724	700	9	441	417	9	169	145
95.0	710	686	97.0	427	403	99.0	156	132
1	695	671	1	413	389	1	142	118
2	681	657	2	399	375	2	129	105
3	666	642	3	385	361	3	116	092
4	652	628	4	372	348	4	103	079
5	638	614	5	358	334	5	089	065
6	623	599	6	344	320	6	076	052
7	609	585	7	331	307	7	063	039
8	595	571	8	317	293	8	050	026
9	581	557	9	303	279	9	037	013
						100.0	024	000

VIII.

ESTIMATION OF PERCENTAGE OF SUGAR BY WEIGHT:

FOR USE WITH SOLUTIONS PREPARED BY ADDITION OF 1/10 VOLUME BASIC ACETATE OF LEAD.

For Soleil-Ventzke Polariscopes.

(SCHMITZ.)

VIII.

Tenths of a Degree.	Per Cent Sucrose.	Polariscope Degrees.	0.5 1.0019	1.0 1.0039	1.5 1.0058	2.0 1.0078	2.5 1.0098	3.0 1.0117	3.5 1.0137	4.0 1.0157	4.5 1.0177
PER CENT BRIX FROM 0.5 TO 12.0.											
0.1°	0.03	1°	0.29	0.29	0.29	0.28	0.28	0.28	0.28	0.28	0.28
0.2	0.06	2		0.57	0.57	0.57	0.57	0.56	0.56	0.56	0.56
0.3	0.08	3		0.85	0.85	0.85	0.85	0.85	0.85	0.84	0.84
0.4	0.11	4			1.14	1.13	1.13	1.13	1.13	1.13	1.12
0.5	0.14	5			1.42	1.42	1.41	1.41	1.41	1.41	1.40
0.6	0.17	6				1.70	1.70	1.69	1.69	1.69	1.68
0.7	0.19	7				1.98	1.98	1.98	1.97	1.97	1.96
0.8	0.22	8					2.26	2.26	2.26	2.25	2.25
0.9	0.25	9						2.54	2.54	2.53	2.53
		10						2.82	2.82	2.81	2.81
		11							3.10	3.09	3.09
		12							3.38	3.38	3.37
		13								3.66	3.65
		14								3.94	3.93
PER CENT BRIX FROM 12.5 TO 20.0.		15									4.21
		16									4.49
		17									
Tenths of a Degree.	Per Cent Sucrose.	18									
		19									
		20									
0.1°	0.03	21									
0.2	0.05	22									
0.3	0.08	23									
0.4	0.11	24									
0.5	0.13	25									
0.6	0.16	26									
0.7	0.19	27									
0.8	0.21	28									
0.9	0.24	29									
		30									
		31									
		32									
		33									
		34									
		35									
		36									
		37									
		38									
		39									

Header span: the columns 0.5 through 4.5 fall under the heading "PER CENT BRIX AND".

Corresponding Specific Gravity.											Polariscope Degrees.
5.0	5.5	6.0	6.5	7.0	7.5	8.0	8.5	9.0	9.5	10.0	
1.0197	1.0217	1.0237	1.0258	1.0278	1.0298	1.0319	1.0339	1.0360	1.0381	1.0401	
0.28	0.28	0.28	0.28	0.28	0.28	0.28	0.28	0.28	0.28	0.28	1°
0.56	0.56	0.56	0.56	0.56	0.55	0.55	0.55	0.55	0.55	0.55	2
0.84	0.84	0.84	0.84	0.83	0.83	0.83	0.83	0.83	0.83	0.82	3
1.12	1.12	1.12	1.11	1.11	1.11	1.11	1.11	1.10	1.10	1.10	4
1.40	1.40	1.40	1.39	1.39	1.39	1.38	1.38	1.38	1.38	1.37	5
1.68	1.68	1.67	1.67	1.67	1.66	1.66	1.66	1.66	1.65	1.65	6
1.96	1.96	1.95	1.95	1.95	1.94	1.94	1.93	1.93	1.93	1.92	7
2.24	2.24	2.23	2.23	2.22	2.22	2.22	2.21	2.21	2.20	2.20	8
2.52	2.52	2.51	2.51	2.50	2.50	2.49	2.49	2.48	2.48	2.47	9
2.80	2.80	2.79	2.79	2.78	2.78	2.77	2.76	2.76	2.75	2.75	10
3.08	3.08	3.07	3.06	3.06	3.05	3.05	3.04	3.03	3.03	3.02	11
3.36	3.36	3.35	3.34	3.34	3.33	3.32	3.32	3.31	3.30	3.30	12
3.64	3.64	3.63	3.62	3.61	3.61	3.60	3.59	3.59	3.58	3.57	13
3.92	3.92	3.91	3.90	3.89	3.88	3.88	3.87	3.86	3.85	3.85	14
4.20	4.19	4.19	4.18	4.17	4.16	4.15	4.15	4.14	4.13	4.12	15
4.48	4.47	4.47	4.46	4.45	4.44	4.43	4.42	4.41	4.40	4.40	16
4.77	4.76	4.75	4.74	4.73	4.72	4.71	4.70	4.69	4.68	4.67	17
	5.03	5.02	5.01	5.00	4.99	4.99	4.97	4.97	4.96	4.95	18
	5.32	5.31	5.29	5.28	5.27	5.26	5.25	5.24	5.23	5.22	19
		5.58	5.57	5.56	5.55	5.54	5.53	5.52	5.51	5.50	20
		5.86	5.85	5.84	5.83	5.82	5.81	5.79	5.78	5.77	21
			6.13	6.12	6.11	6.09	6.08	6.07	6.06	6.05	22
			6.41	6.40	6.38	6.37	6.36	6.35	6.33	6.32	23
				6.67	6.66	6.65	6.64	6.62	6.61	6.60	24
					6.94	6.93	6.91	6.90	6.89	6.87	25
					7.22	7.20	7.19	7.17	7.16	7.15	26
						7.48	7.46	7.45	7.44	7.42	27
						7.76	7.74	7.73	7.71	7.70	28
							8.02	8.00	7.99	7.97	29
								8.28	8.26	8.25	30
								8.55	8.54	8.52	31
								8.83	8.81	8.80	32
									9.09	9.07	33
									9.35		34
										9.62	35
											36
											37
											38
											39

PER CENT BRIX FROM 0.5 TO 12.0		Polari-scope Degrees.	PER CENT BRIX AND								
Tenths of a Degree.	Per Cent Sucrose.		10.5	11.0	11.5	12.0	12.5	13.0	13.5	14.0	14.5
			1.0422	1.0443	1.0464	1.0485	1.0506	1.0528	1.0549	1.0570	1.0592
0.1°	0.03	1°	0.28	0.27	0.27	0.27	0.27	0.27	0.27	0.27	0.27
0.2	0.06	2	0.55	0.55	0.55	0.55	0.54	0.54	0.54	0.54	0.54
0.3	0.08	3	0.82	0.82	0.82	0.82	0.82	0.81	0.81	0.81	0.81
0.4	0.11	4	1.10	1.10	1.09	1.09	1.09	1.09	1.08	1.08	1.08
0.5	0.14	5	1.37	1.37	1.36	1.36	1.36	1.36	1.35	1.35	1.35
0.6	0.17	6	1.64	1.64	1.64	1.64	1.63	1.63	1.62	1.62	1.62
0.7	0.19	7	1.92	1.91	1.91	1.91	1.90	1.90	1.89	1.89	1.89
0.8	0.22	8	2.19	2.19	2.18	2.18	2.18	2.17	2.17	2.16	2.16
0.9	0.25	9	2.47	2.46	2.46	2.45	2.45	2.44	2.44	2.43	2.43
		10	2.74	2.74	2.73	2.73	2.72	2.71	2.71	2.70	2.70
		11	3.02	3.01	3.00	3.00	2.99	2.99	2.98	2.97	2.97
		12	3.29	3.28	3.28	3.27	3.26	3.26	3.25	3.24	3.24
		13	3.56	3.56	3.55	3.54	3.54	3.53	3.52	3.51	3.51
		14	3.84	3.83	3.82	3.82	3.81	3.80	3.79	3.78	3.78
PER CENT BRIX FROM 12.5 TO 20.0.		15	4.11	4.11	4.10	4.09	4.08	4.07	4.06	4.06	4.05
		16	4.39	4.38	4.37	4.36	4.35	4.34	4.33	4.33	4.32
		17	4.66	4.65	4.64	4.63	4.62	4.62	4.61	4.60	4.59
Tenths of a Degree.	Per Cent Sucrose.	18	4.93	4.93	4.91	4.91	4.90	4.89	4.88	4.87	4.86
		19	5.21	5.20	5.19	5.18	5.17	5.16	5.15	5.14	5.13
		20	5.49	5.47	5.46	5.45	5.44	5.43	5.42	5.41	5.40
0.1°	0.03	21	5.76	5.75	5.74	5.73	5.71	5.70	5.69	5.68	5.67
0.2	0.05	22	6.03	6.02	6.01	6.00	5.99	5.97	5.96	5.95	5.94
0.3	0.08	23	6.31	6.30	6.28	6.27	6.26	6.24	6.23	6.22	6.21
0.4	0.11	24	6.58	6.57	6.56	6.54	6.53	6.52	6.50	6.49	6.48
0.5	0.13	25	6.86	6.84	6.83	6.82	6.80	6.79	6.78	6.76	6.75
0.6	0.16	26	7.13	7.12	7.10	7.09	7.07	7.06	7.05	7.03	7.02
0.7	0.19	27	7.41	7.39	7.38	7.36	7.35	7.33	7.32	7.30	7.29
0.8	0.21	28	7.68	7.66	7.65	7.63	7.62	7.60	7.59	7.57	7.56
0.9	0.24	29	7.96	7.94	7.92	7.91	7.89	7.87	7.86	7.84	7.83
		30	8.23	8.21	8.20	8.18	8.16	8.15	8.13	8.11	8.10
		31	8.50	8.49	8.47	8.45	8.44	8.42	8.40	8.39	8.37
		32	8.78	8.76	8.74	8.73	8.71	8.69	8.67	8.66	8.64
		33	9.05	9.03	9.02	9.00	8.98	8.96	8.94	8.93	8.91
		34	9.33	9.31	9.29	9.27	9.25	9.23	9.22	9.20	9.18
		35	9.60	9.58	9.56	9.54	9.53	9.51	9.49	9.47	9.45
		36	9.88	9.86	9.84	9.82	9.80	9.78	9.76	9.74	9.72
		37	10.15	10.13	10.11	10.09	10.07	10.05	10.03	10.01	9.99
		38		10.40	10.38	10.36	10.34	10.32	10.30	10.28	10.26
		39		10.68	10.66	10.64	10.61	10.59	10.57	10.55	10.53

CORRESPONDING SPECIFIC GRAVITY.

15.0	15.5	16.0	16.5	17.0	17.5	18.0	18.5	19.0	19.5	20.0	Polariscope Degrees.
1.0613	1.0635	1.0657	1.0678	1.0700	1.0722	1.0744	1.0766	1.0788	1.0811	1.0833	
0.27	0.27	0.27	0.27	0.27	0.27	0.27	0.27	0.27	0.27	0.26	1°
0.54	0.54	0.54	0.54	0.53	0.53	0.53	0.53	0.53	0.53	0.53	2
0.81	0.81	0.80	0.80	0.80	0.80	0.80	0.80	0.79	0.79	0.79	3
1.08	1.08	1.07	1.07	1.07	1.07	1.06	1.06	1.06	1.06	1.06	4
1.35	1.34	1.34	1.34	1.34	1.33	1.33	1.33	1.32	1.32	1.32	5
1.62	1.61	1.61	1.61	1.60	1.60	1.60	1.59	1.59	1.59	1.58	6
1.88	1.88	1.88	1.87	1.87	1.86	1.86	1.86	1.85	1.85	1.85	7
2.15	2.15	2.15	2.14	2.14	2.13	2.13	2.12	2.12	2.12	2.11	8
2.42	2.42	2.41	2.41	2.40	2.40	2.39	2.39	2.38	2.38	2.37	9
2.69	2.69	2.68	2.68	2.67	2.67	2.66	2.65	2.65	2.64	2.64	10
2.96	2.95	2.95	2.94	2.94	2.93	2.92	2.92	2.91	2.91	2.90	11
3.23	3.22	3.22	3.21	3.20	3.20	3.19	3.18	3.18	3.17	3.17	12
3.50	3.49	3.49	3.48	3.47	3.46	3.46	3.45	3.44	3.44	3.43	13
3.77	3.76	3.75	3.75	3.74	3.73	3.72	3.72	3.71	3.70	3.69	14
4.04	4.03	4.02	4.02	4.01	4.00	3.99	3.98	3.97	3.97	3.96	15
4.31	4.30	4.29	4.28	4.27	4.26	4.26	4.25	4.24	4.23	4.22	16
4.58	4.57	4.56	4.55	4.54	4.53	4.52	4.51	4.50	4.49	4.48	17
4.85	4.84	4.83	4.82	4.81	4.80	4.79	4.78	4.77	4.76	4.75	18
5.12	5.11	5.10	5.09	5.08	5.06	5.05	5.04	5.03	5.02	5.01	19
5.39	5.38	5.36	5.35	5.34	5.33	5.32	5.31	5.30	5.29	5.28	20
5.66	5.65	5.63	5.62	5.61	5.60	5.59	5.58	5.56	5.55	5.54	21
5.93	5.91	5.90	5.89	5.88	5.87	5.85	5.84	5.83	5.82	5.80	22
6.20	6.18	6.17	6.16	6.14	6.13	6.12	6.11	6.09	6.08	6.07	23
6.46	6.45	6.44	6.43	6.41	6.40	6.39	6.37	6.36	6.35	6.33	24
6.73	6.72	6.71	6.69	6.68	6.67	6.65	6.64	6.63	6.61	6.60	25
7.00	6.99	6.97	6.96	6.95	6.93	6.92	6.90	6.89	6.88	6.86	26
7.27	7.26	7.24	7.23	7.21	7.20	7.18	7.17	7.15	7.14	7.13	27
7.54	7.53	7.51	7.50	7.48	7.47	7.45	7.44	7.42	7.40	7.39	28
7.81	7.80	7.78	7.77	7.75	7.73	7.72	7.70	7.68	7.67	7.65	29
8.08	8.06	8.05	8.03	8.02	8.00	7.98	7.97	7.95	7.93	7.92	30
8.35	8.33	8.32	8.30	8.28	8.27	8.25	8.23	8.21	8.20	8.18	31
8.62	8.60	8.58	8.57	8.55	8.53	8.51	8.50	8.48	8.46	8.45	32
8.89	8.87	8.85	8.84	8.82	8.80	8.78	8.76	8.75	8.73	8.71	33
9.16	9.14	9.12	9.10	9.09	9.07	9.05	9.03	9.01	8.99	8.97	34
9.43	9.41	9.39	9.37	9.35	9.34	9.31	9.30	9.28	9.26	9.24	35
9.70	9.68	9.66	9.64	9.62	9.60	9.58	9.56	9.54	9.52	9.50	36
9.97	9.95	9.93	9.91	9.89	9.87	9.85	9.83	9.81	9.79	9.77	37
10.24	10.22	10.20	10.18	10.15	10.13	10.11	10.09	10.07	10.05	10.03	38
10.51	10.49	10.46	10.44	10.42	10.40	10.38	10.36	10.34	10.32	10.29	39

PER CENT BRIX FROM 11.5 TO 22.5 — Tenths of a Degree	Per Cent Sucrose	Polariscope Degrees	11.5 1.0464	12.0 1.0485	12.5 1.0506	13.0 1.0528	13.5 1.0549	14.0 1.0570
		40°	10.93	10.91	10.89	10.86	10.84	10.82
0.1°	0.03	41		11.18	11.16	11.14	11.12	11.09
0.2	0.05	42		11.46	11.43	11.41	11.39	11.36
0.3	0.08	43			11.71	11.68	11.66	11.64
0.4	0.11	44			11.98	11.95	11.93	11.91
0.5	0.13	45			12.25	12.23	12.20	12.18
0.6	0.16	46				12.50	12.47	12.45
0.7	0.19	47					12.74	12.72
0.8	0.21	48					13.02	12.99
0.9	0.24	49						13.26
		50						
		51						
		52						
		53						
		54						
PER CENT BRIX FROM 23.0 TO 24.0 — Tenths of a Degree	Per Cent Sucrose	55						
		56						
		57						
		58						
		59						
		60						
0.1°	0.03	61						
0.2	0.05	62						
0.3	0.08	63						
0.4	0.10	64						
0.5	0.13	65						
0.6	0.16	66						
0.7	0.18	67						
0.8	0.21	68						
0.9	0.23	69						
		70						
		71						
		72						
		73						
		74						
		75						
		76						
		77						
		78						
		79						
		80						

Corresponding Specific Gravity.							Polariscope Degrees.
14.5	15.0	15.5	16.0	16.5	17.0	17.5	
1.0592	1.0613	1.0635	1.0657	1.0678	1.0700	1.0722	
10.80	10.78	10.76	10.73	10.71	10.69	10.67	40
11.07	11.05	11.03	11.00	10.98	10.96	10.94	41
11.34	11.32	11.29	11.27	11.25	11.23	11.20	42
11.61	11.59	11.56	11.54	11.52	11.49	11.47	43
11.88	11.86	11.83	11.81	11.79	11.76	11.74	44
12.15	12.13	12.10	12.08	12.05	12.03	12 01	45
12.42	12.40	12.37	12.35	12.32	12.30	12.27	46
12.69	12.67	12.64	12.61	12.59	12.56	12.54	47
12.97	12.94	12·91	12.88	12.86	12.83	12.81	48
13.23	13.21	13.18	13.15	13.13	13.10	13.07	49
13.50	13.48	13.45	13.42	13.40	13.37	13.34	50
13.78	13.75	13.72	13.69	13.66	13.64	13.61	51
	14.02	13.99	13.96	13.93	13.90	13.88	52
	14.29	14.26	14.23	14.20	14.17	14.14	53
		14.53	14.26	14.50	14.47	14.44	14.41	54
		14.80	14.77	14.74	14.71	14.68	55
			15.03	15.00	14.97	14.94	56
			15.30	15.27	15.24	15.21	57
			15.57	15.54	15.51	15.48	58
				15.81	15.78	15.75	59
					16.05	16.01	60
					16.31	16.28	61
						16.55	62
						16.82	63
							64
							65
							66
							67
							68
							69
							70
							71
							72
							73
							74
							75
							76
							77
							78
							79
							80

| Per Cent Brix from 11.5 to 22.5 | | Polari-scope Degrees | 18.0 | 18.5 | .19.0 | 19.⌄ | 20.0 | 20.5 |
Tenths of a degree.	Per cent Sucrose.		1.0744	1.0766	1.0788	1.0811	1.0833	1.0855
		40°	10.64	10.62	10.60	10.58	10 56	10.54
0.1°	0.03	41	10 91	10.89	10.87	10.85	10.82	10.80
0.2	0.05	42	11.18	11.16	11.13	11.11	11.09	11.07
0.3	0.08	43	11.45	11.42	11.40	11.38	11.35	11.33
0.4	0.11	44	11.71	11.69	11.66	11.64	11.62	11.59
0.5	0.13	45	11.98	11.96	11.93	11.91	11.88	11.86
0.6	0.16	46	12.25	12.22	12.20	12.17	12.15	12.12
0.7	0.19	47	12.51	12.49	12.46	12.44	12.41	12.39
0.8	0.21	48	12.78	12.75	12.73	12.70	12.67	12.65
0.9	0.24	49	13.05	13.02	12.99	12.97	12.94	12.91
		50	13.31	13.29	13.26	13.23	13.20	13.18
		51	13.58	13.55	13.52	13.50	13.47	13.44
		52	13.85	13.82	13.79	13.76	13.73	13.70
		53	14.11	14.08	14.05	14.03	14.00	13.97
		54	14.38	14.35	14.32	14.29	14.26	14.23
Per Cent Brix from 23.0 to 24.0.		55	14.65	14.62	14.59	14.56	14.53	14.50
		56	14.91	14.88	14.85	14.82	14.79	14.76
		57	15.18	15.15	15.12	15.09	15.06	15.02
Tenths of a degree.	Per cent Sucrose.	58	15.45	15.42	15.38	15.35	15.32	15.29
		59	15.71	15.68	15.65	15.62	15.58	15.55
		60	15.98	15.95	15.92	15.88	15.85	15.82
0.1°	0.03	61	16.25	16.21	16.18	16.15	16.11	16.08
0.2	0.05	62	16.52	16.48	16.45	16 41	16.38	16.35
0.3	0.08	63	16.78	16.75	16.71	16.68	16.64	16.61
0.4	0.10	64	17.05	17.01	16.98	16.94	16.91	16.87
0.5	0.13	65	17.32	17.28	17.24	17.21	17.17	17.14
0.6	0.16	66		17.55	17.51	17.47	17.44	17.40
0.7	0.18	67		17.81	17.78	17.74	17.70	17.67
0.8	0.21	68			18.04	18.00	17.97	17.93
0.9	0.23	69			18.31	18.27	18.23	18.19
		70				18.53	18.50	18.46
		71					18.76	18.72
		72					19.03	18.99
		73						19.25
		74						19.52
		75						19.78
		76						
		77						
		78						
		79						
		80						

CORRESPONDING SPECIFIC GRAVITY.							Polariscope Degrees.
21 0	21.5	22.0	22.5	23.0	23.5	24.0	
1.0878	1.0900	1.0923	1.0946	1.0969	1.0992	1.1015	
10.52	10.49	10.47	10.45	10.43	10.41	10.38	40°
10.78	10.76	10.74	10.71	10.69	10.67	10.65	41
11.04	11.02	11.00	10.97	10.95	10.93	10.90	42
11.31	11.28	11.26	11.24	11.21	11.19	11.17	43
11.57	11.55	11.52	11.50	11.47	11.45	11.42	44
11.83	11.81	11.78	11.76	11.73	11.71	11.69	45
12.09	12.07	12 05	12.02	12.00	11.97	11.94	46
12.36	12.33	12.31	12.28	12.26	12.23	12.21	47
12.62	12.60	12.57	12.54	12.52	12.49	12.47	48
12.88	12.86	12.83	12.81	12.78	12.75	12.73	49
13.15	13.12	13.09	13.07	13.04	13.01	12.99	50
13.41	13.39	13.36	13.33	13.30	13.27	13.25	51
13.68	13.65	13.62	13.59	13.56	13.53	13.51	52
13.94	13.91	13.88	13.85	13.82	13.79	13.77	53
14.20	14.17	14.14	14.11	14.08	14.06	14.02	54
14.47	14.44	14.41	14.38	14.35	14.32	14.29	55
14.73	14.70	14.67	14.64	14.61	14.58	14.55	56
14.99	14.96	14.93	14.90	14.87	14.84	14.81	57
15.26	15.23	15.19	15.16	15.13	15.10	15.07	58
15.52	15.49	15.46	15.42	15.39	15.36	15.33	59
15.78	15.75	15.72	15.69	15.65	15.62	15.59	60
16.05	16.01	15.98	15.95	15.91	15.88	15.85	61
16.31	16.28	16.24	16.21	16.18	16.14	16.11	62
16.57	16.54	16.51	16.47	16.44	16.40	16.37	63
16.84	16.80	16.77	16.73	16.70	16.66	16.63	64
17.10	17.07	17.03	17.00	16.96	16.92	16.89	65
17.37	17.33	17.29	17.26	17.22	17.19	17.15	66
17.63	17.59	17.56	17.52	17.48	17.45	17.41	67
17.89	17.86	17.82	17.78	17.74	17.71	17.67	68
18.16	18.12	18.08	18.04	18.00	17.97	17.93	69
18.42	18.38	18.35	18.31	18.27	18.23	18.19	70
18.68	18.65	18.61	18.57	18.53	18.49	18.45	71
18.95	18.91	18.87	18.83	18.79	18.75	18.71	72
19.21	19.17	19.13	19.09	19.05	19.01	18.97	73
19.48	19.44	19.40	19.35	19.31	19.27	19.23	74
19.74	19.70	19.66	19.62	19.57	19.53	19.49	75
20.00	19.96	19.92	19.88	19.84	19.80	19.75	76
20.27	20.22	20.18	20.14	20.10	20.06	20.01	77
	20.49	20.45	20.40	20.36	20.32	20.27	78
	20.75	20.71	20.66	20.62	20.58	20.54	79
		20.97	20.93	20.88	20.84	20.80	80

IX.

POUNDS SOLIDS PER CUBIC FOOT IN SUGAR SOLUTIONS.

Calculated for Wiechmann: Sugar Analysis, from the following data taken from Everett: Physical Units and Constants. 2d edition 1886.

1 cubic centimetre of water at 17.5° C. weighs 0.9987605 grms.
1 cubic foot = 28316 cubic centimetres.
1 kilogramme = 2.2046212 lbs.
Hence 1 cubic foot of water at 17.5° C. weighs 62.3487 lbs.

FORMULÆ.

I.   62.3487 × Specific Gravity of Sugar Solution.

II. $\dfrac{\text{Result obtained by I.} \times \text{Degree Brix}}{100}$

= Pounds Solids per Cubic Foot.

SUGAR ANALYSIS.

IX.

Degree Baumé.	Degree Brix.	Specific Gravity.	Lbs. solids in 1 cu. ft.	Degree Baumé.	Degree Brix.	Specific Gravity.	Lbs. solids in 1 cu. ft.
0.0	0.0	1.00000	0.000	26.5	47.7	1.22019	36.289
0.5	0.9	1.00349	0.563	27.0	48.7	1.22564	37.215
1.0	1.8	1.00701	1.130	27.5	49.6	1.23058	38.056
1.5	2.6	1.01015	1.638	28.0	50.5	1.23555	38.903
2.0	3.5	1.01371	2.212	28.5	51.5	1.24111	39.852
2.5	4.4	1.01730	2.791	29 0	52.4	1.24614	40.712
3.0	5.3	1.02091	3.374	29.5	53.4	1.25177	41.677
3.5	6.2	1.02454	3.900	30.0	54.3	1.25687	42.552
4.0	7.0	1.02779	4.486	30.5	55.2	1.26200	43.434
4.5	7.9	1.03146	5.081	31.0	56.2	1.26773	44.421
5.0	8.8	1.03517	5.680	31.5	57.2	1.27351	45.418
5.5	9.7	1.03889	6.283	32.0	58.1	1.27874	46.322
6.0	10.6	1.04264	6.891	32.5	59.1	1.28459	47.335
6.5	11.5	1.04641	7.503	33.0	60.0	1.28989	48.254
7.0	12.4	1.05021	8.119	33.5	61.0	1.29581	49.283
7.5	13.2	1.05361	8.671	34.0	61.9	1.30117	50.217
8.0	14.1	1.05746	9.296	34.5	62.9	1.30717	51.264
8.5	15.0	1.06133	9.926	35.0	63.9	1.31320	52.319
9.0	15.9	1.06522	10.560	35.5	64.9	1.31928	53.384
9.5	16.8	1.06914	11.199	36.0	65.8	1.32478	54.350
10.0	17.7	1.07309	11.842	36.5	66.8	1.33093	55.432
10.5	18.6	1.07706	12.491	37.0	67.8	1.33712	56.523
11.0	19.5	1.08106	13.144	37.5	68.8	1.34335	57.624
11.5	20.4	1.08509	13.801	38.0	69.8	1.34962	58.735
12.0	21.3	1.08914	14.464	38.5	70.7	1.35530	59.742
12.5	22.2	1.09321	15.132	39.0	71.7	1.36164	60.871
13.0	23.1	1.09732	15.804	39.5	72.7	1.36803	62.009
13.5	24.0	1.10145	16.482	40.0	73.7	1.37446	63.158
14.0	24.9	1.10560	17.164	40.5	74.7	1.38092	64.316
14.5	25.8	1.10979	17.852	41.0	75.7	1.38743	65.484
15.0	26.7	1.11400	18.545	41.5	76.7	1.39397	66.662
15.5	27.6	1.11824	19.243	42.0	77.7	1.40056	67.850
16.0	28.5	1.12250	19.946	42.5	78.8	1.40785	69.169
16.5	29.4	1.12679	20.655	43.0	79.8	1.41452	70.378
17.0	30.3	1.13111	21.369	43.5	80.8	1.42123	71.598
17.5	31.2	1.13545	22.088	44.0	81.8	1.42798	72.829
18.0	32.1	1.13983	22.812	44.5	82.8	1.43478	74.070
18.5	33.0	1.14423	23.543	45.0	83.9	1.44229	75.447
19.0	33.9	1.14866	24.278	45.5	84.9	1.44917	76.710
19.5	34.8	1.15312	25.020	46.0	85.9	1.45609	77.985
20.0	35.7	1.15760	25.766	46.5	87.0	1.46374	79.398
20.5	36.6	1.16212	26.519	47.0	88.0	1.47074	80.695
21.0	37.6	1.16717	27.362	47.5	89.1	1.47849	82.134
21.5	38.5	1.17174	28.127	48.0	90.1	1.48558	83.454
22.0	39.4	1.17635	28.897	48.5	91.2	1.49342	84.919
22.5	40.3	1.18098	29.674	49.0	92.3	1.50130	86.397
23.0	41.2	1.18564	30.456	49.5	93.3	1.50852	87.753
23.5	42.2	1.19086	31.333	50.0	94.4	1.51649	89.256
24.0	43.1	1.19558	32.128	50.5	95.5	1.52449	90.773
24.5	44.0	1.20033	32.929	51.0	96.6	1.53254	92.303
25.0	44.9	1.20565	33.737	51.5	97.7	1.54068	93.850
25.5	45.9	1.21046	34.641	52.0	98.8	1.54890	95.413
26.0	46.8	1.21531	35.462	52.5	99.9	1.55711	96.987

X.

FACTORS FOR THE CALCULATION OF CLERGET INVERSIONS.

Calculated for Wiechmann: Sugar Analysis, by the formula:

$$\text{Factor} = \frac{100}{142.66 - \dfrac{t}{2}}.$$

X.

Temperature.	Factor.	Temperature.	Factor.
10°	0.7257	21°	0.7567
11	0.7291	22	0.7595
12	0.7317	23	0.7624
13	0.7344	24	0.7653
14	0.7371	25	0.7683
15	0.7397	26	0.7712
16	0.7426	27	0.7742
17	0.7454·	28	0.7772
18	0.7482	29	0.7802
19	0.7510	30	0.7833
20	0.7538		

XI.

DETERMINATION OF TOTAL SUGAR.

German Government: Law of July 9, 1887.

XI.

Mgr. Sucrose.	Mgr. Copper.	Mgr. Sucrose.	Mgr. Copper.	Mgr. Sucrose.	Mgr. Copper.	Mgr. Sucrose.	Mgr. Copper.
40	79.0	73	145.2	106	208.6	139	269.1
41	81.0	74	147.1	107	210.5	140	270.9
42	83.0	75	149.1	108	212.3	141	272.7
43	85.2	76	151.0	109	214.2	142	274.5
44	87.2	77	153.0	110	216.1	143	276.3
45	89.2	78	155.0	111	217.9	144	278.1
46	91.2	79	156.9	112	219.8	145	279.9
47	93.3	80	158.9	113	221.6	146	281.6
48	95.3	81	160.8	114	223.5	147	283.4
49	97.3	82	162.8	115	225.3	148	285.2
50	99.3	83	164.7	116	227.2	149	286.9
51	101.3	84	166.6	117	229.0	150	288.8
52	103.3	85	168.6	118	230.9	151	290.5
53	105.3	86	170.5	119	232.8	152	292.3
54	107.3	87	172.4	120	234.6	153	294.0
55	109.4	88	174.3	121	236.4	154	295.7
56	111.4	89	176.3	122	238.3	155	297.5
57	113.4	90	178.2	123	240.2	156	299.2
58	115.4	91	180.1	124	242.0	157	300.9
59	117.4	92	182.0	125	243.9	158	302.6
60	119.5	93	183.9	126	245.7	159	304.4
61	121.5	94	185.8	127	247.5	160	306.1
62	123.5	95	187.8	128	249.3	161	307.8
63	125.4	96	189.7	129	251.2	162	309.5
64	127.4	97	191.6	130	252.9	163	311.3
65	129.4	98	193.5	131	254.7	164	313.0
66	131.4	99	195.4	132	256.5	165	314.7
67	133.4	100	197.3	133	258.3	166	316.4
68	135.3	101	199.2	134	260.1	167	318.1
69	137.3	102	201.1	135	261.9	168	319.9
70	139.3	103	202.9	136	263.7	169	321.6
71	141.3	104	204.8	137	265.5	170	323.3
72	143.2	105	206.7	138	267.3		

XII.

DETERMINATION OF INVERT-SUGAR.
VOLUMETRIC METHOD.

(Using Fehling's Solution.)

5 grammes to 100 cubic centimetres.

Divide 1.00 by the number of cubic centimetres used of above solution, and multiply result by 100.

XII.

Number of c.c. used.	Per cent of Invert-Sugar.	Number of c.c. used.	Per cent of Invert-Sugar.	Number of c.c. used.	Per cent of Invert-Sugar.	Number of c.c. used.	Per cent of Invert-Sugar.
1	100.00	26	3.85	51	1.96	76	1.32
2	50.00	27	3.70	52	1.92	77	1.30
3	33.33	28	3.57	53	1.89	78	1.28
4	25.00	29	3.45	54	1.85	79	1.27
5	20.00	30	3.33	55	1.82	80	1.25
6	16.67	31	3.23	56	1.79	81	1.23
7	14.29	32	3.13	57	1.75	82	1.22
8	12.50	33	3.03	58	1.72	83	1.20
9	11.11	34	2.94	59	1.69	84	1.19
10	10.00	35	2.86	60	1.67	85	1.18
11	9.09	36	2.78	61	1.64	86	1.16
12	8.33	37	2.70	62	1.61	87	1.15
13	7.69	38	2.63	63	1.59	88	1.14
14	7.14	39	2.56	64	1.56	89	1.12
15	6.67	40	2.50	65	1.54	90	1.11
16	6.25	41	2.44	66	1.52	91	1.10
17	5.88	42	2.38	67	1.49	92	1.09
18	5.55	43	2.33	68	1.47	93	1.08
19	5.26	44	2.27	69	1.45	94	1.06
20	5.00	45	2.22	70	1.43	95	1.05
21	4.76	46	2.17	71	1.41	96	1.04
22	4.55	47	2.13	72	1.39	97	1.03
23	4.35	48	2.08	73	1.37	98	1.02
24	4.17	49	2.04	74	1.35	99	1.01
25	4.00	50	2.00	75	1.33	100	1.00

XIII.

DETERMINATION OF INVERT-SUGAR.
GRAVIMETRIC METHOD.

(Using Fehling's Solution.)

HERZFELD, HILLER, MEISSL.

SUGAR ANALYSIS.

XIII.

$R:I.$	$Z = 200$ mg.	175 mg.	150 mg.	125 mg.	100 mg.	75 mg.	50 mg.
0 : 100	56.4	55.4	54.5	53.8	53.2	53.0	53.0
10 : 90	56.3	55.3	54.4	53.8	53.2	52.9	52.9
20 : 80	56.2	55.2	54.3	53.7	53.2	52.7	52.7
30 : 70	56.1	55.1	54.2	53.7	53.2	52.6	52.6
40 : 60	55.9	55.0	54.1	53.6	53.1	52.5	52.4
50 : 50	55.7	54.9	54.0	53.5	53.1	52.3	52.2
60 : 40	55.6	54.7	53.8	53.2	52.8	52.1	51.9
70 : 30	55.5	54.5	53.5	52.9	52.5	51.9	51.6
80 : 20	55.4	54.3	53.3	52.7	52.2	51.7	51.3
90 : 10	54.6	53.6	53.1	52.6	52.1	51.6	51.2
91 : 9	54.1	53.6	52.6	52.1	51.6	51.2	50.7
92 : 8	53.6	53.1	52.1	51.6	51.2	50.7	50.3
93 : 7	53.6	53.1	52.1	51.2	50.7	50.3	49.8
94 : 6	53.1	52.6	51.6	50.7	50.3	49.8	48.9
95 : 5	52.6	52.1	51.2	50.3	49.4	48.9	48.5
96 : 4	52.1	51.2	50.7	49.8	48.9	47.7	46.9
97 : 3	50.7	50.3	49.8	48.9	47.7	46.2	45.1
98 : 2	49.9	48.9	48.5	47.3	45.8	43.3	40.0
99 : 1	47.7	47.3	46.5	45.1	43.3	41.2	38.1

XIV.

DETERMINATION OF INVERT-SUGAR.
GRAVIMETRIC METHOD.

(Using Soldaini's Solution.)

PREUSS.

XIV.

Mgr. Invert-Sugar.	Mgr. Copper.	Mgr. Invert-Sugar.	Mgr. Copper.	Mgr. Invert-Sugar.	Mgr. Copper.
5	18.8	23	76.0	41	130.7
6	21.9	24	79.1	42	133.6
7	25.2	25	82.2	43	136.5
8	28.4	26	85.3	44	139.5
9	31.6	27	88.5	45	142.4
10	34.9	28	91.4	46	145.4
11	38.1	29	94.5	47	148.3
12	41.3	30	97.6	48	151.2
13	44.5	31	100.6	49	154.1
14	47.7	32	103.6	50	157.0
15	50.9	33	106.6	55	171.3
16	54.0	34	109.7	60	185.5
17	57.2	35	112.7	65	200.4
18	60.3	36	115.7	70	213.1
19	63.5	37	118.7	75	226.6
20	66.6	38	121.8	80	240.0
21	69.7	39	124.8		
22	72.9	40	127.8		

XV.

DETERMINATION OF DEXTROSE.

From E. Wein, Tabellen zur Quantitativen Bestimmung der Zuckerarten.

F. ALLIHN.

XV.

Mgr. Copper.	Mgr. Dextrose.	Mgr. Copper.	Mgr. Dextrose.	Mgr. Copper.	Mgr. Dextrose.	Mgr. Copper.	Mgr. Dextrose.
10	6.1	58	29.8	106	54.0	154	78.6
11	6.6	59	30.3	107	54.5	155	79.1
12	7.1	60	30.8	108	55.0	156	79.6
13	7.6	61	31.3	109	55.5	157	80.1
14	8.1	62	31.8	110	56.0	158	80.7
15	8.6	63	32.3	111	56.5	159	81.2
16	9.0	64	32.8	112	57.0	160	81.7
17	9.5	65	33.3	113	57.5	161	82.2
18	10.0	66	33.8	114	58.0	162	82.7
19	10.5	67	34.3	115	58.6	163	83.3
20	11.0	68	34.8	116	59.1	164	83.8
21	11.5	69	35.3	117	59.6	165	84.3
22	12.0	70	35.8	118	60.1	166	84.8
23	12.5	71	36.3	119	60.6	167	85.3
24	13.0	72	36.8	120	61.1	168	85.9
25	13.5	73	37.3	121	61.6	169	86.4
26	14.0	74	37.8	122	62.1	170	86.9
27	14.5	75	38.3	123	62.6	171	87.4
28	15.0	76	38.8	124	63.1	172	87.9
29	15.5	77	39.3	125	63.7	173	88.5
30	16.0	78	39.8	126	64.2	174	89.0
31	16.5	79	40.3	127	64.7	175	89.5
32	17.0	80	40.8	128	65.2	176	90.0
33	17.5	81	41.3	129	65.7	177	90.5
34	18.0	82	41.8	130	66.2	178	91.1
35	18.5	83	42.3	131	66.7	179	91.6
36	18.9	84	42.8	132	67.2	180	92.1
37	19.4	85	43.4	133	67.7	181	92.6
38	19.9	86	43.9	134	68.2	182	93.1
39	20.4	87	44.4	135	68.8	183	93.7
40	20.9	88	44.9	136	69.3	184	94.2
41	21.4	89	45.4	137	69.8	185	94.7
42	21.9	90	45.9	138	70.3	186	95.2
43	22.4	91	46.4	139	70.8	187	95.7
44	22.9	92	46.9	140	71.3	188	96.3
45	23.4	93	47.4	141	71.8	189	96.8
46	23.9	94	47.9	142	72.3	190	97.3
47	24.4	95	48.4	143	72.9	191	97.8
48	24.9	96	48.9	144	73.4	192	98.4
49	25.4	97	49.4	145	73.9	193	98.9
50	25.9	98	49.9	146	74.4	194	99.4
51	26.4	99	50.4	147	74.9	195	100.0
52	26.9	100	50.9	148	75.5	196	100.5
53	27.4	101	51.4	149	76.0	197	101.0
54	27.9	102	51.9	150	76.5	198	101.5
55	28.4	103	52.4	151	77.0	199	102.0
56	28.8	104	52.9	152	77.5	200	102.6
57	29.3	105	53.5	153	78.1	201	103.2

Mgr. Copper.	Mgr. Dextrose.	Mgr. Copper.	Mgr. Dextrose.	Mgr. Copper.	Mgr. Dextrose.	Mgr. Copper.	Mgr. Dextrose.
202	103.7	250	129.2	298	155.4	346	182.1
203	104.2	251	129.7	299	156.0	347	182.6
204	104.7	252	130.3	300	156.5	348	183.2
205	105.3	253	130.8	301	157.1	349	183.7
206	105.8	254	131.4	302	157.6	350	184.3
207	106.3	255	131.9	303	158.2	351	184.9
208	106.8	256	132.4	304	158.7	352	185.4
209	107.4	257	133.0	305	159.3	353	186.0
210	107.9	258	133.5	306	159.8	354	186.6
211	108.4	259	134.1	307	160.4	355	187.2
212	109.0	260	134.6	308	160.9	356	187.7
213	109.5	261	135.1	309	161.5	357	188.3
214	110.0	262	135.7	310	162.0	358	188.9
215	110.6	263	136.2	311	162.6	359	189.4
216	111.1	264	136.8	312	163.1	360	190.0
217	111.6	265	137.3	313	163.7	361	190.6
218	112.1	266	137.8	314	164.2	362	191.1
219	112.7	267	138.4	315	164.8	363	191.7
220	113.2	268	138.9	316	165.3	364	192.3
221	113.7	269	139.5	317	165.9	365	192.9
222	114.3	270	140.0	318	166.4	366	193.4
223	114.8	271	140.6	319	167.0	367	194.0
224	115.3	272	141.1	320	167.5	368	194.6
225	115.9	273	141.7	321	168.1	369	195.1
226	116.4	274	142.2	322	168.6	370	195.7
227	116.9	275	142.8	323	169.2	371	196.3
228	117.4	276	143.3	324	169.7	372	196.8
229	118.0	277	143.9	325	170.3	373	197.4
230	118.5	278	144.4	326	170.9	374	198.0
231	119.0	279	145.0	327	171.4	375	198.6
232	119.6	280	145.5	328	172.0	376	199.1
233	120.1	281	146.1	329	172.5	377	199.7
234	120.7	282	146.6	330	173.1	378	200.3
235	121.2	283	147.2	331	173.7	379	200.8
236	121.7	284	147.7	332	174.2	380	201.4
237	122.3	285	148.3	333	174.8	381	202.0
238	122.8	286	148.8	334	175.3	382	202.5
239	123.4	287	149.4	335	175.9	383	203.1
240	123.9	288	149.9	336	176.5	384	203.7
241	124.4	289	150.5	337	177.0	385	204.3
242	125.0	290	151.0	338	177.6	386	204.8
243	125.5	291	151.6	339	178.1	387	205.4
244	126.0	292	152.1	340	178.7	388	206.0
245	126.6	293	152.7	341	179.3	389	206.5
246	127.1	294	153.2	342	179.8	390	207.1
247	127.6	295	153.8	343	180.4	391	207.7
248	128.1	296	154.3	344	180.9	392	208.3
249	128.7	297	154.9	345	181.5	393	208.8

Mgr. Copper.	Mgr. Dextrose.	Mgr. Copper.	Mgr. Dextrose.	Mgr. Copper.	Mgr. Dextrose.	Mgr. Copper.	Mgr. Dextrose.
394	209.4	412	219.9	430	230.4	447	240.4
395	210.0	413	220.4	431	231.0	448	241.0
396	210.6	414	221.0	432	231.6	449	241.6
397	211.2	415	221.6	433	232.2	450	242.2
398	211.7	416	222.2	434	232.8	451	242.8
399	212.3	417	222.8	435	233.4	452	243.4
400	212.9	418	223.3	436	233.9	453	244.0
401	213.5	419	223.9	437	234.5	454	244.6
402	214.1	420	224.5	438	235.1	455	245.2
403	214.6	421	225.1	439	235.7	456	245.7
404	215.2	422	225.7	440	236.3	457	246.3
405	215.8	423	226.3	441	236.9	458	246.9
406	216.4	424	226.9	442	237.5	459	247.5
407	217.0	425	227.5	443	238.1	460	248.1
408	217.5	426	228.0	444	238.7	461	248.7
409	218.1	427	228.6	445	239.3	462	249.3
410	218.7	428	229.2	446	239.8	463	249.9
411	219.3	429	229.8				

XVI.

DETERMINATION OF LÆVULOSE.

From E. Wein, Tabellen zur Quantitativen Bestimmung der Zuckerarten.

LEHMANN.

XVI.

Mgr. Copper.	Mgr. Lævulose.	Mgr. Copper.	Mgr. Lævulose.	Mgr. Copper.	Mgr. Lævulose.	Mgr. Copper.	Mgr. Lævulose.
20	7.15	68	35.21	116	64.21	164	94.17
21	7.78	69	35.81	117	64.84	165	94.80
22	8.41	70	36.40	118	65.46	166	95.44
23	9.04	71	37.00	119	66.09	167	96.08
24	9.67	72	37.59	120	66.72	168	96.71
25	10.30	73	38.19	121	67.32	169	97.35
26	10.81	74	38.78	122	67.92	170	97.99
27	11.33	75	39.38	123	68.53	171	98.63
28	11.84	76	39.98	124	69.13	172	99.27
29	12.36	77	40.58	125	69.73	173	99.90
30	12.87	78	41.17	126	70.35	174	100.54
31	13.46	79	41.77	127	70.96	175	101.18
32	14.05	80	42.37	128	71.58	176	101.82
33	14.64	81	42.97	129	72.19	177	102.46
34	15.23	82	43.57	130	72.81	178	103.11
35	15.82	83	44.16	131	73.43	179	103.75
36	16.40	84	44.76	132	74.05	180	104.39
37	16.99	85	45.36	133	74.67	181	105.04
38	17.57	86	45.96	134	75.29	182	105.68
39	18.16	87	46.57	135	75.91	183	106.33
40	18.74	88	47.17	136	76.53	184	106.97
41	19.32	89	47.78	137	77.15	185	107.62
42	19.91	90	48.38	138	77.77	186	108.27
43	20.49	91	48.98	139	78.39	187	108.92
44	21.08	92	49.58	140	79.01	188	109.56
45	21.66	93	50.18	141	79.64	189	110.21
46	22.25	94	50.78	142	80.28	190	110.86
47	22.83	95	51.38	143	80.91	191	111.50
48	23.42	96	51.98	144	81.55	192	112.14
49	24.00	97	52.58	145	82.18	193	112.78
50	24.59	98	53.19	146	82.81	194	113.42
51	25.18	99	53.79	147	83.43	195	114.06
52	25.76	100	54.39	148	84.06	196	114.72
53	26.35	101	55.00	149	84.68	197	115.38
54	26.93	102	55.62	150	85.31	198	116.04
55	27.52	103	56.23	151	85.93	199	116.70
56	28.11	104	56.85	152	86.55	200	117.36
57	28.70	105	57.46	153	87.16	201	118.02
58	29.30	106	58.07	154	87.78	202	118.68
59	29.89	107	58.68	155	88.40	203	119.33
60	30.48	108	59.30	156	89.05	204	119.99
61	31.07	109	59.91	157	89.69	205	120.65
62	31.66	110	60.52	158	90.34	206	121.30
63	32.25	111	61.13	159	90.98	207	121.96
64	32.84	112	61.74	160	91.63	208	122.61
65	33.43	113	62.36	161	92.26	209	123.27
66	34.02	114	62.97	162	92.90	210	123.92
67	34.62	115	63.58	163	93.53	211	124.58

Mgr. Copper.	Mgr. Lævulose.	Mgr. Copper.	Mgr. Lævulose.	Mgr. Copper.	Mgr. Lævulose.	Mgr. Copper.	Mgr. Lævulose.
212	125.24	256	154.91	300	185.63	343	216.97
213	125.90	257	155.65	301	186.35	344	217.72
214	126.56	258	156.40	302	187.06	345	218.47
215	127.22	259	157.14	303	187.78	346	219.21
216	127.85	260	157.88	304	188.49	347	219.97
217	128.48	261	158.49	305	189.21	348	220.71
218	129.10	262	159.09	306	189.93	349	221.46
219	129.73	263	159.70	307	190.65	350	222.21
220	130.36	264	160.30	308	191.37	351	222.96
221	131.07	265	160.91	309	192.09	352	223.72
222	131.77	266	161.63	310	192.81	353	224.47
223	132.48	267	162.35	311	193.53	354	225.23
224	133.18	268	163.07	312	194.25	355	225.98
225	133.89	269	163.79	313	194.97	356	226.74
226	134.56	270	164.51	314	195.69	357	227.49
227	135.23	271	165.21	315	196.41	358	228.25
228	135.89	272	165.90	316	197.12	359	229.00
229	136.89	273	166.60	317	197.83	360	229.76
230	137.23	274	167.29	318	198.55	361	230.52
231	137.90	275	167.99	319	199.26	362	231.28
232	138.57	276	168.68	320	199.97	363	232.05
233	139.25	277	169.37	321	200.71	364	232.81
234	139.92	278	170.06	322	201.44	365	233.57
235	140.59	279	170.75	323	202.18	366	234.33
236	141.27	280	171.44	324	202.91	367	235.10
237	141.94	281	172.14	325	203.65	368	235.86
238	142.62	282	172.85	326	204.39	369	236.63
239	143.29	283	173.55	327	205.13	370	237.39
240	143.97	284	174.26	328	205.88	371	238.16
241	144.65	285	174.96	329	206.62	372	238.93
242	145.32	286	175.67	330	207.36	373	239.69
243	146.00	287	176.39	331	208.10	374	240.46
244	146.67	288	177.10	332	208.83	375	241.23
245	147.35	289	177.82	333	209.57	376	241.87
246	148.03	290	178.53	334	210.30	377	242.51
247	148.71	291	179.24	335	211.04	378	243.15
248	149.40	292	179.95	336	211.78	379	243.79
249	150.08	293	180.65	337	212.52	380	244.43
250	150.76	294	181.36	338	213.25	381	245.34
251	151.44	295	182.07	339	213.99	382	246.25
252	152.12	296	182.78	340	214.73	383	247.17
253	152.81	297	183.49	341	215.48	384	248.08
254	153.49	298	184.21	342	216.23	385	248.99
255	154.17	299	184.92				

XVII.

DENSITY OF WATER AT THE TEMPERATURES FROM 0° TO 50° CENTIGRADE, RELATIVE TO ITS DENSITY AT 4° CENTIGRADE.

ROSETTI.

Based on results obtained by Kopp, Despretz, Hagen, Matthiessen, Rosetti.

XVII.

Temperature: Degrees Centigrade.	Density of Water relative to its Density at 4° C.	Temperature: Degrees Centigrade.	Density of Water relative to its Density at 4° C.
0°	0.99987	25°	0.99712
1	0.99993	26	0.99687
2	0.99997	27	0.99660
3	0.99999	28	0.99633
4	1.00000	29	0.99605
5	0.99999	30	0.99577
6	0.99997	31	0.99547
7	0.99993	32	0.99517
8	0.99989	33	0.99485
9	0.99982	34	0.99452
10	0.99975	35	0.99418
11	0.99966	36	0.99383
12	0.99955	37	0.99347
13	0.99943	38	0.99310
14	0.99930	39	0.99273
15	0.99916	40	0.99235
16	0.99900	41	0.99197
17	0.99884	42	0.99158
18	0.99865	43	0.99118
19	0.99846	44	0.99078
20	0.99826	45	0.99037
21	0.99805	46	0.98996
22	0.99783	47	0.98954
23	0.99760	48	0.98910
24	0.99737	49	0.98865
		50	0.98819

XVIII.

COMPARISON OF THERMOMETRIC SCALES.

$$F = \frac{9}{5} C + 32 = \frac{9}{4} R + 32.$$

$$C = \frac{5}{9} (F - 32) = \frac{5}{4} R.$$

$$R = \frac{4}{9} (F - 32) = \frac{4}{5} C.$$

XVIII.
CENTIGRADE, FAHRENHEIT, RÉAUMUR.

Centi-grade.	Fahren-heit.	Réaumur.	Centi-grade.	Fahren-heit.	Réaumur.	Centi-grade.	Fahren-heit.	Réaumur.
100	212	80	53	127.4	42.4	6	42.8	4.8
99	210.2	79.2	52	125.6	41.6	5	41	4
98	208.4	78.4	51	123.8	40.8	4	39.2	3.2
97	206.6	77.6	50	122	40	3	37.4	2.4
96	204.8	76.8	49	120.2	39.2	2	35.6	1.6
95	203	76	48	118.4	38.4	1	33.8	0.8
94	201.2	75.2	47	116.6	37.6	0	32	0
93	199.4	74.4	46	114.8	36.8	−1	30.2	−0.8
92	197.6	73.6	45	113	36	−2	28.4	−1.6
91	195.8	72.8	44	111.2	35.2	−3	26.6	−2.4
90	194	72	43	109.4	34.4	−4	24.8	−3.2
89	192.2	71.2	42	107.6	33.6	−5	23	−4
88	190.4	70.4	41	105.8	32.8	−6	21.2	−4.8
87	188.6	69.6	40	104	32	−7	19.4	−5.6
86	186.8	68.8	39	102.2	31.2	−8	17.6	−6.4
85	185	68	38	100.4	30.4	−9	15.8	−7.2
84	183.2	67.2	37	98.6	29.6	−10	14	−8
83	181.4	66.4	36	96.8	28.8	−11	12.2	−8.8
82	179.6	65.6	35	95	28	−12	10.4	−9.6
81	177.8	64.8	34	93.2	27.2	−13	8.6	−10.4
80	176	64	33	91.4	26.4	−14	6.8	−11.2
79	174.2	63.2	32	89.6	25.6	−15	5	−12
78	172.4	62.4	31	87.8	24.8	−16	3.2	−12.8
77	170.6	61.6	30	86	24	−17	1.4	−13.6
76	168.8	60.8	29	84.2	23.2	−18	0.4	−14.4
75	167	60	28	82.4	22.4	−19	−2.2	−15.2
74	165.2	59.2	27	80.6	21.6	−20	−4.	−16
73	163.4	58.4	26	78.8	20.8	−21	−5.8	−16.8
72	161.6	57.6	25	77	20	−22	−7.6	−17.6
71	159.8	56.8	24	75.2	19.2	−23	−9.4	−18.4
70	158	56	23	73.4	18.4	−24	−11.2	−19.2
69	156.2	55.2	22	71.6	17.6	−25	−13.	−20
68	154.4	54.4	21	69.8	16.8	−26	−14.8	−20.8
67	152.6	53.6	20	68	16	−27	−16.6	−21.6
66	150.8	52.8	19	66.2	15.2	−28	−18.4	−22.4
65	149	52	18	64.4	14.4	−29	−20.2	−23.2
64	147.2	51.2	17	62.6	13.6	−30	−22	−24
63	145.4	50.4	16	60.8	12.8	−31	−23.8	−24.8
62	143.6	49.6	15	59	12	−32	−25.6	−25.6
61	141.8	48.8	14	57.2	11.2	−33	−27.4	−26.4
60	140	48	13	55.4	10.4	−34	−29.2	−27.2
59	138.2	47.2	12	53.6	9.6	−35	−31	−28
58	136.4	46.4	11	51.8	8.8	−36	−32.8	−28.8
57	134.6	45.6	10	50	8	−37	−34.6	−29.6
56	132.8	44.8	9	48.2	7.2	−38	−36.4	−30.4
55	131	44	8	46.4	6.4	−39	−38.2	−31.2
54	129.2	43.2	7	44.6	5.6	−40	−40	−32

XVIII.
FAHRENHEIT, CENTIGRADE, RÉAUMUR.

Fahrenheit.	Centigrade.	Réaumur.	Fahrenheit.	Centigrade.	Réaumur.	Fahrenheit.	Centigrade.	Réaumur.
°	°	°	°	°	°	°	°	°
212	100	80	165	73.89	59.11	118	47.78	38.22
211	99.44	79.56	164	73.33	58.67	117	47.22	37.78
210	98.89	79.11	163	72.78	58.22	116	46.67	37.33
209	98.33	78.67	162	72.22	57.78	115	46.11	36.89
208	97.78	78.22	161	71.67	57.33	114	45.55	36.44
207	97.22	77.78	160	71.11	56.89	113	45	36
206	96.67	77.33	159	70.55	56.44	112	44.44	35.56
205	96.11	76.89	158	70	56	111	43.89	35.11
204	95.55	76.44	157	69.44	55.56	110	43.33	34.67
203	95	76	156	68.89	55.11	109	42.78	34.22
202	94.44	75.56	155	68.33	54.67	108	42.22	33.78
201	93.89	75.11	154	67.78	54.22	107	41.67	33.33
200	93.33	74.67	153	67.22	53.78	106	41.11	32.89
199	92.78	74.22	152	66.67	53.33	105	40.55	32.44
198	92.22	73.78	151	66.11	52.89	104	40	32
197	91.67	73.33	150	65.55	52.44	103	39.44	31.56
196	91.11	72.89	149	65	52	102	38.89	31.11
195	90.55	72.44	148	64.44	51.56	101	38.33	30.67
194	90	72	147	63.89	51.11	100	37.78	30.22
193	89.44	71.56	146	63.33	50.67	99	37.22	29.78
192	88.89	71.11	145	62.78	50.22	98	36.67	29.33
191	88.33	70.67	144	62.22	49.78	97	36.11	28.89
190	87.78	70.22	143	61.67	49.33	96	35.55	28.44
189	87.22	69.78	142	61.11	48.89	95	35	28
188	86.67	69.33	141	60.55	48.44	94	34.44	27.56
187	86.11	68.89	140	60	48	93	33.89	27.11
186	85.55	68.44	139	59.44	47.56	92	33.33	26.67
185	85	68	138	58.89	47.11	91	32.78	26.22
184	84.44	67.56	137	58.33	46.67	90	32.22	25.78
183	83.89	67.11	136	57.78	46.22	89	31.67	25.33
182	83.33	66.67	135	57.22	45.78	88	31.11	24.89
181	82.78	66.22	134	56.67	45.33	87	30.55	24.44
180	82.22	65.78	133	56.11	44.89	86	30	24
179	81.67	65.33	132	55.55	44.44	85	29.44	23.56
178	81.11	64.89	131	55	44	84	28.89	23.11
177	80.55	64.44	130	54.44	43.56	83	28.33	22.67
176	80	64	129	53.89	43.11	82	27.78	22.22
175	79.44	63.56	128	53.33	42.67	81	27.22	21.78
174	78.89	63.11	127	52.78	42.22	80	26.67	21.33
173	78.33	62.67	126	52.22	41.78	79	26.11	20.89
172	77.78	62.22	125	51.67	41.33	78	25.55	20.44
171	77.22	61.78	124	51.11	40.89	77	25	20
170	76.67	61.33	123	50.55	40.44	76	24.44	19.56
169	76.11°	60.89	122	50	40	75	23.89	19.11
168	75.55	60.44	121	49.44	39.56	74	23.33	18.67
167	75	60	120	48.89	39.11	73	22.78	18.22
166	74.44	59.56	119	48.33	38.67	72	22.22	17.78

Fahrenheit.	Centigrade.	Réaumur.	Fahrenheit.	Centigrade.	Réaumur.	Fahrenheit.	Centigrade.	Réaumur.
°	°	°	°	°	°	°	°	°
71	21.67	17.33	33	0.55	0.44	—4	—20	—16
70	21.11	16.89	32	0	0	—5	—20.55	—16.44
69	20.55	16.44	31	—0.55	—0.44	—6	—21.11	—16.89
68	20	16	30	—1.11	—0.89	—7	—21.67	—17.33
67	19.44	15.56	29	—1.67	—1.33	—8	—22.22	—17.78
66	18.89	15.11	28	—2.22	—1.78	—9	—22.78	—18.22
65	18.33	14.67	27	—2.78	—2.22	—10	—23.33	—18.67
64	17.78	14.22	26	—3.33	—2.67	—11	—23.89	—19.11
63	17.22	13.78	25	—3.89	—3.11	—12	—24.44	—19.56
62	16.67	13.33	24	—4.44	—3.56	—13	—25	—20
61	16.11	12.89	23	—5	—4	—14	—25.55	—20.44
60	15.55	12.44	22	—5.55	—4.44	—15	—26.11	—20.89
59	15	12	21	—6.11	—4.89	—16	—26.67	—21.33
58	14.44	11.56	20	—6.67	—5.33	—17	—27.22	—21.78
57	13.89	11.11	19	—7.22	—5.78	—18	—27.78	—22.22
56	13.33	10.67	18	—7.78	—6.22	—19	—28.33	—22.67
55	12.78	10.22	17	—8.33	—6.67	—20	—28.89	—23.11
54	12.22	9.78	16	—8.89	—7.11	—21	—29.44	—23.56
53	11.67	9.33	15	—9.44	—7.56	—22	—30	—24
52	11.11	8.89	14	—10	—8	—23	—30.55	—24.44
51	10.55	8.44	13	—10.55	—8.44	—24	—31.11	—24.89
50	10	8	12	—11.11	—8.89	—25	—31.67	—25.33
49	9.44	7.56	11	—11.67	—9.33	—26	—32.22	—25.78
48	8.89	7.11	10	—12.22	—9.78	—27	—32.78	—26.22
47	8.33	6.67	9	—12.78	—10.22	—28	—33.33	—26.67
46	7.78	6.22	8	—13.33	—10.67	—29	—33.89	—27.11
45	7.22	5.78	7	—13.89	—11.11	—30	—34.44	—27.56
44	6.67	5.33	6	—14.44	—11.56	—31	—35	—28
43	6.11	4.89	5	—15	—12	—32	—35.55	—28.44
42	5.55	4.44	4	—15.55	—12.44	—33	—36.11	—28.89
41	5	4	3	—16.11	—12.89	—34	—36.67	—29.33
40	4.44	3.56	2	—16.67	—13.33	—35	—37.22	—29.78
39	3.89	3.11	1	—17.22	—13.78	—36	—37.78	—30.22
38	3.33	2.67	0	—17.78	—14.22	—37	—38.33	—30.67
37	2.78	2.22	—1	—18.33	—14.67	—38	—38.89	—31.11
36	2.22	1.78	—2	—18.89	—15.11	—39	—39.44	—31.56
35	1.67	1.33	—3	—19.44	—15.56	—40	—40	—32
34	1.11	0.89						

XIX.

TABLES FOR CONVERTING CUSTOMARY AND METRIC WEIGHTS AND MEASURES.

UNITED STATES COAST AND GEODETIC SURVEY.

OFFICE OF STANDARD WEIGHTS AND MEASURES.

T. C. MENDENHALL, Superintendent.

WASHINGTON, D.C., 1890.

———

[*Authorized Reprint.*]

CUSTOMARY TO METRIC.

LINEAR.

	Inches to millimetres.	Feet to metres.	Yards to metres.	Miles to kilometres.
1 =	25.4000	0.304801	0.914402	1.60935
2 =	50.8001	0.609601	1.828804	3.21869
3 =	76.2001	0.914402	2.743205	4.82804
4 =	101.6002	1.219202	3.657607	6.43739
5 =	127.0002	1.524003	4.572009	8.04674
6 =	152.4003	1.828804	5.486411	9.65608
7 =	177.8003	2.133604	6.400813	11.26543
8 =	203.2004	2.438405	7.315215	12.87478
9 =	228.6004	2.743205	8.229616	14.48412

CAPACITY.

	Fluid drams to millilitres or cubic centimetres.	Fluid ounces to millilitres.	Quarts to litres.	Gallons to litres.
1 =	3.70	29.57	0.94636	3.78544
2 =	7.39	59.15	1.89272	7.57088
3 =	11.09	88.72	2.83908	11.35632
4 =	14.79	118.30	3.78544	15.14176
5 =	18.48	147.87	4.73180	18.92720
6 =	22.18	177.44	5.67816	22.71264
7 =	25.88	207.02	6.62452	26.49808
8 =	29.57	236.59	7.57088	30.28352
9 =	33.28	266.16	8.51724	34.06896

SQUARE.

	Square inches to square centimetres.	Square feet to square decimetres.	Square yards to square metres.	Acres to hectares.
1 =	6.452	9 290	0.836	0.4047
2 =	12.903	18.581	1.672	0.8094
3 =	19.355	27.871	2.508	1.2141
4 =	25.807	37.161	3.344	1.6187
5 =	32.258	46.452	4.181	2.0234
6 =	38.710	55.742	5.017	2.4281
7 =	45.161	65.032	5.853	2.8328
8 =	51.613	74.323	6.689	3.2375
9 =	58.065	83.613	7.525	3.6422

WEIGHT.

	Grains to milligrammes.	Avoirdupois ounces to grammes.	Avoirdupois pounds to kilogrammes.	Troy ounces to grammes.
1 =	64.7989	28.3495	0.45359	31.10348
2 =	129.5978	56.6991	0.90719	62.20696
3 =	194.3968	85.0486	1.36078	93.31044
4 =	259.1957	113.3981	1.81437	124.41392
5 =	323.9946	141.7476	2.26796	155.51740
6 =	388.7935	170.0972	2.72156	186.62089
7 =	453.5924	198.4467	3.17515	217.72437
8 =	518.3914	226.7962	3.62874	248.82785
9 =	583.1903	255.1457	4.08233	279.93133

CUBIC.

	Cubic inches to cubic centimetres.	Cubic feet to cubic metres.	Cubic yards to cubic metres.	Bushels to hectolitres.
1 =	16.387	0.02832	0.765	0.35242
2 =	32.774	0.05663	1.529	0.70485
3 =	49.161	0.08495	2.294	1.05727
4 =	65.549	0.11327	3.058	1.40969
5 =	81.936	0.14158	3.823	1.76211
6 =	98.323	0.16990	4.587	2.11454
7 =	114.710	0.19822	5.352	2.46696
8 =	131.097	0.22654	6.116	2.81938
9 =	147.484	0.25485	6.881	3.17181

1 chain	=	20.1169 metres.
1 square mile	=	259 hectares.
1 fathom	=	1.829 metres.
1 nautical mile	=	1853.27 metres.
1 foot = 0 304801 metre,		9.4840158 log.
1 avoir. pound =		453.5924277 gram.
15432.35639 grains	=	1 kilogramme.

METRIC TO CUSTOMARY.

LINEAR.

	Metres to inches.	Metres to feet.	Metres to yards.	Kilometres to miles.
1 =	39.3700	3.28083	1.093611	0.62137
2 =	78.7400	6.56167	2.187222	1.24274
3 =	118.1100	9.84250	3.280833	1.86411
4 =	157.4800	13.12333	4.374444	2.48548
5 =	196.8500	16.40417	5.468056	3.10685
6 =	236.2200	19.68500	6.561667	3.72822
7 =	275.5900	22.96583	7.655278	4.34959
8 =	314.9600	26.24667	8.748889	4.97096
9 =	354.3300	29.52750	9.842500	5.59233

CAPACITY.

	Millilitres or cubic centimetres to fluid drams.	Centilitres to fluid ounces.	Litres to quarts.	Dekalitres to gallons.	Hektolitres to bushels
1 =	0.27	0.338	1.0567	2.6417	2.8375
2 =	0.54	0.676	2.1134	5.2834	5.6750
3 =	0.81	1.014	3.1700	7.9251	8.5125
4 =	1.08	1.352	4.2267	10.5668	11.3500
5 =	1.35	1.691	5.2834	13.2085	14.1875
6 =	1.62	2.029	6.3401	15.8502	17.0250
7 =	1.89	2.368	7.3968	18.4919	19.8625
8 =	2.16	2.706	8.4534	21.1336	22.7000
9 =	2.43	3.043	9.5101	23.7753	25.5375

SQUARE.

	Square centimetres to square inches.	Square metres to square feet.	Square metres to square yards.	Hectares to acres.
1 =	0.1550	10.764	1.196	2.471
2 =	0.3100	21.528	2.392	4.942
3 =	0.4650	32.292	3.588	7.413
4 =	0.6200	43.055	4.784	9.884
5 =	0.7750	53.819	5.980	12.355
6 =	0.9300	64.583	7.176	14.826
7 =	1.0850	75.347	8.372	17.297
8 =	1.2400	86.111	9.568	19.768
9 =	1.3950	96.874	10.764	22.239

WEIGHT.

	Milligrammes to grains.	Kilogrammes to grains.	Hectogrammes (100 grammes) to ounces av.	Kilogrammes to pounds avoirdupois.
1 =	0.01543	15432.36	3.5274	2.20462
2 =	0.03086	30864.71	7.0548	4.40924
3 =	0.04630	46297.07	10.5822	6.61386
4 =	0.06173	61729.43	14.1096	8.81849
5 =	0.07716	77161.78	17.6370	11.02311
6 =	0.09259	92594.14	21.1644	13.22773
7 =	0.10803	108026.49	24.6918	15.43235
8 =	0.12346	123458.85	28.2192	17.63697
9 =	0.13889	138891.21	31.7466	19.84159

CUBIC.

	Cubic centimetres to cubic inches.	Cubic decimetres to cubic inches.	Cubic metres to cubic feet.	Cubic metres to cubic yards.
1 =	0.0610	61.023	35.314	1.308
2 =	0.1220	122.047	70.629	2.616
3 =	0.1831	183.070	105.943	3.924
4 =	0.2441	244.093	141.258	5.232
5 =	0.3051	305.117	176.572	6.540
6 =	0.3661	366.140	211.887	7.848
7 =	0.4272	427.163	247.201	9.156
8 =	0.4882	488.187	282.516	10.464
9 =	0.5492	549.210	317.830	11.771

WEIGHT.—(Continued.)

	Quintals to pounds av.	Milliers or tonnes to pounds av.	Grammes to ounces, Troy.
1 =	220.46	2204.6	0.03215
2 =	440.92	4409.2	0.06430
3 =	661.38	6613.8	0.09645
4 =	881.84	8818.4	0.12860
5 =	1102.30	11023.0	0.16075
6 =	1322.76	13227.6	0.19290
7 =	1543.22	15432.2	0.22505
8 =	1763.68	17636.8	0.25721
9 =	1984.14	19841.4	0.28936

INDEX.

184   INDEX.

S